T0331888

Mathematical Formulas for Industrial and Mechanical Engineering

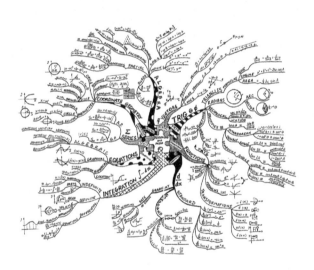

Mathematical Formulas for Industrial and Mechanical Engineering

Seifedine Kadry

American University of the Middle East, Kuwait

AMSTERDAM • BOSTON • HEIDELBERG • LONDON • NEW YORK • OXFORD
ELSEVIER PARIS • SAN DIEGO • SAN FRANCISCO • SINGAPORE • SYDNEY • TOKYO

Elsevier
32 Jamestown Road, London NW1 7BY
225 Wyman Street, Waltham, MA 02451, USA

Copyright © 2014 Elsevier Inc. All rights reserved

No part of this publication may be reproduced or transmitted in any form or by any means,
electronic or mechanical, including photocopying, recording, or any information storage
and retrieval system, without permission in writing from the publisher. Details on how to
seek permission, further information about the Publisher's permissions policies and our
arrangement with organizations such as the Copyright Clearance Center and the Copyright
Licensing Agency, can be found at our website: www.elsevier.com/permissions

This book and the individual contributions contained in it are protected under copyright by
the Publisher (other than as may be noted herein).

Notices
Knowledge and best practice in this field are constantly changing. As new research and
experience broaden our understanding, changes in research methods, professional practices,
or medical treatment may become necessary.

Practitioners and researchers must always rely on their own experience and knowledge in
evaluating and using any information, methods, compounds, or experiments described
herein. In using such information or methods they should be mindful of their own safety
and the safety of others, including parties for whom they have a professional responsibility.

To the fullest extent of the law, neither the Publisher nor the authors, contributors, or editors,
assume any liability for any injury and/or damage to persons or property as a matter of
products liability, negligence or otherwise, or from any use or operation of any methods,
products, instructions, or ideas contained in the material herein.

British Library Cataloguing-in-Publication Data
A catalogue record for this book is available from the British Library

Library of Congress Cataloging-in-Publication Data
A catalog record for this book is available from the Library of Congress

ISBN: 978-0-12-420131-6

For information on all Elsevier publications
visit our website at store.elsevier.com

This book has been manufactured using Print On Demand technology. Each copy is
produced to order and is limited to black ink. The online version of this book will show
color figures where appropriate.

Working together
to grow libraries in
developing countries

www.elsevier.com • www.bookaid.org

Contents

Preface

The material of this book has been compiled so that it serves the needs of students and teachers as well as professional workers who use mathematics. The contents and size make it especially convenient and portable. The widespread availability and low price of scientific calculators have greatly reduced the need for many numerical tables that make most handbooks bulky. However, most calculators do not give integrals, derivatives, series, and other mathematical formulas and figures that are often needed. Accordingly, this book contains that information in an easy way to access in addition to illustrative examples that make formulas more clear. To facilitate the use of this book, the author and publisher have worked together to make the format attractive and clear. Students and professionals alike will find this book a valuable supplement to standard textbooks, a source for review, and a handy reference for many years.

Biography

Seifedine Kadry is an associate professor of Applied Mathematics in the American University of the Middle East Kuwait. He received his Masters degree in Modelling and Intensive Computing (2001) from the Lebanese University—EPFL-INRIA. He did his doctoral research (2003—2007) in applied mathematics from Blaise Pascal University, Clermont Ferrand II, France. He worked as Head of Software Support and Analysis Unit of First National Bank where he designed and implemented the data warehouse and business intelligence; he has published one book and more than 50 papers on applied maths, computer science, and stochastic systems in peer-reviewed journals.

1 Symbols and Special Numbers

In this chapter, several symbols used in mathematics are defined. Some special numbers are given with examples and many conversion formulas are studied. This chapter is essential to understand the next chapters. Topics discussed in this chapter are as follows:

- Basic mathematical symbols
- Base algebra symbols
- Linear algebra symbols
- Probability and statistics symbols
- Geometry symbols
- Set theory symbols
- Logic symbols
- Calculus symbols
- Numeral symbols
- Greek alphabet letters
- Roman numerals
- Special numbers like prime numbers
- Conversion formulas
- Basic area, perimeter, and volume formulas.

Students encounter many mathematical symbols during their math courses. The following sections show a categorical list of the math symbols, how to read them, and some examples.

1.1 Basic Mathematical Symbols

Symbols	How to Read It	How to Use It	Examples
$=$	equals	equality	$3 + 7 = 10$
$\neq$	does not equal	inequality	$12 \neq 10$
$<>$			
$! =$			
$<$	is less than (strict)	less than	$2 < 5$
$\leq$	is less than or equal	less than or equal to	$12 <= 12$
$<=$			
$>$	is greater than (strict)	greater than	$7 > 3$

(Continued)

Mathematical Formulas for Industrial and Mechanical Engineering. DOI: http://dx.doi.org/10.1016/B978-0-12-420131-6.00001-4
© 2014 Elsevier Inc. All rights reserved.

(Continued)

Symbols	How to Read It	How to Use It	Examples
$\geq$ >=	is greater than or equal	greater than or equal to	$15 >= 15$
[ ]	brackets	calculate expression inside first	$[(1 + 2) * (1 + 5)] = 18$
( )	parentheses	calculate expression inside first	$2 \times (3 + 5) = 16$
+	plus		$7 + 8 = 15$
−	minus		$8 - 2 = 6$
$\times$	times	multiplication	$7 \times 8 = 56$
•	dot		
*	asterisk		
$\div$	division sign	divided by	$12/3 = 4$
/	division slash		
---	horizontal line		
$\pm$	plus-minus	both plus and minus operations	$3 \pm 5 = 8$ and -2
$\mp$	minus-plus	both minus and plus operations	$3 \mp 5 = -2$ and 8
$\sqrt{\ }$	square root		$\sqrt{9} = 3$
$3\sqrt{a}$	cube root		$3\sqrt{8} = 2$
$n\sqrt{a}$	nth root (radical)		for $n = 3$, $n\sqrt{8} = 2$
$\lvert\ldots\rvert$	absolute value or modulus		$\lvert-5\rvert = \lvert5\rvert$ (absolute value) $\lvert3 + 4i\rvert = 5$ (modulus of complex number)
$\vert$	divides		$5\vert20$
$\Rightarrow$	implies		$x = 2 \Rightarrow x2 = 4$
$\Leftrightarrow$	equivalence		$x + 5 = y + 2 \Leftrightarrow x + 3 = y$
$\forall$	for each		
$\exists$	there exists		
$\exists!$	there exists. exactly one		
.	period	decimal point, decimal separator	$2.56 = 2 + 56/100$
a^b	power	exponent	$2^3 = 8$
$a^\wedge b$	caret	exponent	$2^\wedge 3 = 8$
%	percent	$1\% = 1/100$	$10\% \times 30 = 3$
‰	per mille	$1‰ = 1/1000 = 0.1\%$	$10‰ \times 30 = 0.3$

1.2 Basic Algebra Symbols

Symbols	How to Read It	How to Use It	Examples
X	x variable	unknown value to find	when $2x = 4$, then $x = 2$
$\equiv$	equivalence	identical to	
$\triangleq$	equal by definition	equal by definition	
$:=$	definition		$\cosh x := (1/2)(\exp x + \exp(-x))$
$\approx$	approximately equal	approximation	$\pi \approx 3.14159$
$\sim$	approximately equal	weak approximation	$11 \sim 10$
$\propto$	proportional	is proportional to	if $y = 5x$, then $y \propto x$
∞	lemniscates	infinity symbol	
$\ll$	is much less than	is much less than	$3 \ll 1000$
$\gg$	is much greater than	is much greater than	$95 \gg 0.2$
$\lfloor x \rfloor$	floor brackets	rounds number to lower integer	$\lfloor 4.3 \rfloor = 4$
$\lceil x \rceil$	ceiling brackets	rounds number to upper integer	$\lceil 4.3 \rceil = 5$
$x!$	exclamation mark	factorial	$4! = 1*2*3*4 = 24$
$f(x)$	function of x	maps values of x to $f(x)$	$f(x) = 3x + 5$
$(f \circ g)$	function composition	$(f \circ g)(x) = f(g(x))$	$f(x) = 3x$, $g(x) = x - 1 \Rightarrow (f \circ g)(x) = 3(x - 1)$
(a,b)	open interval	$(a,b) \triangleq \{x \mid a < x < b\}$	$x \in (2,6)$
$[a,b]$	closed interval	$[a,b] \triangleq \{x \mid a \le x \le b\}$	$x \in [2,6]$
Δ	delta	change/difference	
Δ	discriminant	$\Delta = b^2 - 4ac$	
$\sum$	sigma	summation—sum of all values in range of series	$\displaystyle\sum_{n=2}^{5} n^2 = 2^2 + 3^2 + 4^2 + 5^2 = 54$
$\sum\sum$	sigma	double summation	$\displaystyle\sum_{j=1}^{2}\sum_{i=1}^{8} x_{i,j} = \sum_{i=1}^{8} x_{i,1} + \sum_{i=1}^{8} x_{i,2}$
Π	capital pi	product—product of all values in range of series	$\Pi x_i = x_1, \cdot x_2, \cdot \ldots \cdot x_n$
E	e constant/Euler's number	$e = 2.718281828\ldots$	
Γ	Euler−Mascheroni constant	$\gamma = 0.527721566\ldots$	
Φ	golden ratio	$\varphi = 1.61803398875$	

1.3 Linear Algebra Symbol

Symbols	How to Read It	How to Use It	Examples
·	dot	scalar product	$a \cdot b$
×	cross	vector product	$a \times b$
$A \otimes B$	tensor product	tensor product of A and B	$A \otimes B$
$\langle x, y \rangle$	inner product		
[ ]	brackets	matrix of numbers	
( )	parentheses	matrix of numbers	
$|A|$	determinant	determinant of matrix A	
$\det(A)$	determinant	determinant of matrix A	
$\|x\|$	double vertical bars	norm	
A^{T}	transpose	matrix transpose	$(A^{\mathrm{T}})_{ij} = (A)_{ji}$
$A^{\dagger}$	Hermitian matrix	matrix conjugate transpose	$(A^{\dagger})_{ij} = (A)_{ji}$
A^{*}	Hermitian matrix	matrix conjugate transpose	$(A^{*})_{ij} = (A)_{ji}$
A^{-1}	inverse matrix	$A A^{-1} = I$	
$\mathrm{rank}(A)$	matrix rank	rank of matrix A	$\mathrm{rank}(A) = 3$
$\dim(U)$	dimension	dimension of matrix A	$\mathrm{rank}(U) = 3$

1.4 Probability and Statistics Symbols

Symbols	How to Read It	How to Use It	Examples
$P(A)$	probability function	probability of event A	$P(A) = 0.5$
$P(A \cap B)$	probability of events intersection	probability that of events A and B	$P(A \cap B) = 0.5$
$P(A \cup B)$	probability of events union	probability that of events A or B	$P(A \cup B) = 0.5$
$P(A|B)$	conditional probability function	probability of event A given event B occurred	$P(A|B) = 0.3$
$f(x)$	probability density function (pdf)	$P(a \leq x \leq b) = \int f(x) \mathrm{d}x$	
$F(x)$	cumulative distribution function (cdf)	$F(x) = P(X \leq x)$	
μ	population mean	mean of population values	$\mu = 10$
$E(X)$	expectation value	expected value of random variable X	$E(X) = 10$
$E(X|Y)$	conditional expectation	expected value of random variable X given Y	$E(X|Y = 2) = 5$
$\mathrm{var}(X)$	variance	variance of random variable X	$\mathrm{var}(X) = 4$

(Continued)

(Continued)

Symbols	How to Read It	How to Use It	Examples
σ^2	variance	variance of population values	$\sigma^2 = 4$
std(X)	standard deviation	standard deviation of random variable X	std(X) = 2
σ_X	standard deviation	standard deviation value of random variable X	$\sigma_X = 2$
$\tilde{x}$	median	middle value of random variable x	$\tilde{x} = 5$
cov(X,Y)	covariance	covariance of random variables X and Y	cov(X,Y) = 4
corr(X,Y)	correlation	correlation of random variables X and Y	corr(X,Y) = 3
$\rho_{X,Y}$	correlation	correlation of random variables X and Y	$\rho_{X,Y} = 3$
Mod	mode	value that occurs most frequently in population	
MR	mid range	$MR = (x_{max} + x_{min})/2$	
Md	sample median	half the population is below this value	
Q_1	lower/first quartile	25% of population are below this value	
Q_2	median/second quartile	50% of population are below this value = median of samples	
Q_3	upper/third quartile	75% of population are below this value	
X	sample mean	average/arithmetic mean	$x = (2 + 5 + 9)/3 = 5.333$
s^2	sample variance	population samples variance estimator	$s^2 = 4$
S	sample standard deviation	population samples standard deviation estimator	$s = 2$
z_x	standard score	$z_x = (x - x)/s_x$	
$X\sim$	distribution of X	distribution of random variable X	$X \sim N(0,3)$
$N(\mu,\sigma^2)$	normal distribution	Gaussian distribution	$X \sim N(0,3)$
$U(a,b)$	uniform distribution	equal probability in range a, b	$X \sim U(0,3)$
exp(λ)	exponential distribution	$f(x) = \lambda e^{-\lambda x}, x \geq 0$	

(Continued)

(Continued)

Symbols	How to Read It	How to Use It	Examples
gamma(c, λ)	gamma distribution	$f(x) = \lambda c x^{c-1} e^{-\lambda x}/\Gamma(c)$, $x \geq 0$	
$\chi^2(k)$	chi-square distribution	$f(x) = x^{k/2-1} e^{-x/2}/(2^{k/2}\Gamma(k/2))$	
$F(k_1, k_2)$	F distribution		
Bin(n,p)	binomial distribution	$f(k) = {}_nC_k p^k (1-p)^{n-k}$	
Poisson(λ)	Poisson distribution	$f(k) = \lambda^k e^{-\lambda}/k!$	
Geom(p)	geometric distribution	$f(k) = p(1-p)^k$	
HG(N,K,n)	hyper-geometric distribution		
Bern(p)	Bernoulli distribution		
$n!$	factorial	$n! = 1 \cdot 2 \cdot 3 \cdots n$	$5! = 1 \cdot 2 \cdot 3 \cdot 4 \cdot 5 = 120$
${}_nP_k$	permutation	${}_nP_k = \frac{n!}{(n-k)!}$	${}_5P_3 = 5!/(5-3)! = 60$
${}_nC_k \binom{n}{k}$	combination	${}_nC_k = \binom{n}{k} = \frac{n!}{k!(n-k)!}$	${}_5C_3 = 5!/[3!(5-3)!] = 10$

1.5 Geometry Symbols

Symbols	How to Read It	How to Use It	Examples
$\angle$	angle	formed by two rays	$\angle ABC = 30°$
$\measuredangle$	measured angle		$\measuredangle ABC = 30°$
$\sphericalangle$	spherical angle		$\sphericalangle AOB = 30°$
$\llcorner$	right angle	$= 90°$	$\alpha = 90°$
$°$	degree	1 turn $= 360°$	$\alpha = 60°$
$'$	arcminute	$1° = 60'$	$\alpha = 60°59'$
$''$	arcsecond	$1' = 60''$	$\alpha = 60°59'59''$
$\overleftrightarrow{AB}$	line	infinite line	
AB	line segment	line from point A to point B	
$\overrightarrow{AB}$	ray	line that start from point A	
$\overset{\frown}{AB}$	arc	arc from point A to point B	
$\mid$	perpendicular	perpendicular lines (90° angle)	$AC\mid BC$
$\parallel$	parallel	parallel lines	$AB\parallel CD$
$\cong$	congruent to	equivalence of geometric shapes and size	$\triangle ABC \cong \triangle XYZ$
$\sim$	similarity	same shapes, not same size	$\triangle ABC \sim \triangle XYZ$
$\triangle$	triangle	triangle shape	$\triangle ABC \cong \triangle BCD$
$\lvert x-y \rvert$	distance	distance between points x and y	$\lvert x-y \rvert = 5$

(Continued)

Symbols	How to Read It	How to Use It	Examples
π	pi constant	$\pi = 3.141592654\ldots$ is the ratio between the circumference and diameter of a circle	$c = \pi \cdot d = 2 \cdot \pi \cdot r$
rad	radians	radians angle unit	$360° = 2\pi$ rad
grad	grads	grads angle unit	$360° = 400$ grad

1.6 Set Theory Symbols

Symbols	How to Read It	How to Use It	Examples
{ }	set	a collection of elements	$A = \{3,7,9,14\}$, $B = \{9,14,28\}$
$A \cap B$	intersection	objects that belong to set A and set B	$A \cap B = \{9,14\}$
$A \cup B$	union	objects that belong to set A or set B	$A \cup B = \{3,7,9,14,28\}$
$A \subseteq B$	subset	subset has less elements or equal to the set	$\{9,14,28\} \subseteq \{9,14,28\}$
$A \subset B$	proper subset/strict subset	subset has less elements than the set	$\{9,14\} \subset \{9,14,28\}$
$A \not\subset B$	not subset	left set not a subset of right set	$\{9,66\} \not\subset \{9,14,28\}$
$A \supseteq B$	superset	set A has more elements or equal to the set B	$\{9,14,28\} \supseteq \{9,14,28\}$
$A \supset B$	proper superset/strict superset	set A has more elements than set B	$\{9,14,28\} \supset \{9,14\}$
$A \not\supset B$	not superset	set A is not a superset of set B	$\{9,14,28\} \not\supset \{9,66\}$
2^A	power set	all subsets of A	
$\mathsf{P}(A)$	power set	all subsets of A	
$A = B$	equality	both sets have the same members	$A = \{3,9,14\}$, $B = \{3,9,14\}$, $A = B$
A^c	complement	all the objects that do not belong to set A	
$A \backslash B$	relative complement	objects that belong to A and not to B	$A = \{3,9,14\}$, $B = \{1,2,3\}$, $A - B = \{9,14\}$
$A - B$	relative complement	objects that belong to A and not to B	$A = \{3,9,14\}$, $B = \{1,2,3\}$, $A - B = \{9,14\}$
$A \triangle B$	symmetric difference	objects that belong to A or B but not to their intersection	$A = \{3,9,14\}$, $B = \{1,2,3\}$, $A \triangle B = \{1,2,9,14\}$

(Continued)

(Continued)

Symbols	How to Read It	How to Use It	Examples
$A \ominus B$	symmetric difference	objects that belong to A or B but not to their intersection	$A = \{3,9,14\}$, $B = \{1,2,3\}$, $A \ominus B = \{1,2,9,14\}$
$a \in A$	element of	set membership	$A = \{3,9,14\}$, $3 \in A$
$x \notin A$	not element of	no set membership	$A = \{3,9,14\}$, $1 \notin A$
(a,b)	ordered pair	collection of two elements	
$A \times B$	Cartesian product	set of all ordered pairs from A and B	
$\lvert A \rvert$	cardinality	the number of elements of set A	$A = \{3,9,14\}$, $\lvert A \rvert = 3$
$\#A$	cardinality	the number of elements of set A	$A = \{3,9,14\}$, $\#A = 3$
$\aleph_0$	aleph-null	infinite cardinality of natural numbers set	
$\aleph_1$	aleph-one	cardinality of countable ordinal numbers set	
$\varnothing$	empty set	$\varnothing = \{\ \}$	$C = \{\varnothing\}$
U	universal set	set of all possible values	
N_0	natural numbers/whole numbers set (with zero)	$N_0 = \{0,1,2,3,4,\ldots\}$	$0 \in N_0$
N_1	natural numbers/whole numbers set (without zero)	$N_1 = \{1,2,3,4,5,\ldots\}$	$6 \in N_1$
Z	integer numbers set	$Z = \{\ldots -3, -2, -1, 0,1,2,3,\ldots\}$	$-6 \in Z$
Q	rational numbers set	$Q = \{x \mid x = a/b,\ a,b \in N\}$	$2/6 \in Q$
R	real numbers set	$R = \{x \mid -\infty < x < \infty\}$	$6.343434 \in R$
C	complex numbers set	$C = \{z \mid z = a + bi,\ -\infty < a < \infty,\ -\infty < b < \infty\}$	$6 + 2i \in C$

1.7 Logic Symbols

Symbols	How to Read It	How to Use It	Examples
·	and	and	$x \cdot y$
∧	caret/circumflex	and	$x {\wedge} y$
&	ampersand	and	$x\ \&\ y$
+	plus	or	$x + y$

(Continued)

Symbols	How to Read It	How to Use It	Examples
∨	reversed caret	or	$x \vee y$
|	vertical line	or	$x\|y$
x'	single quote	not—negation	x'
x	bar	not—negation	X
¬	not	not—negation	$\neg x$
!	exclamation mark	not—negation	$!x$
⊕	circled plus/oplus	exclusive or—xor	$x \oplus y$
~	tilde	negation	$\sim x$
⇒	implies		
⇔	equivalent	if and only if	
∀	for all	.	
∃	there exists		
∄	there does not exists		
∴	therefore		
∵	because/since		

1.8 Calculus Symbols

Symbols	How to Read It	How to Use It	Examples
$\lim_{x \to x0} f(x)$	Limit	limit value of a function	
ε	epsilon	represents a very small number, near zero	$\varepsilon \to 0$
e	e constant/Euler's number	$e = 2.718281828\ldots$	$e = \lim(1 + 1/x)^x, x \to \infty$
y'	derivative	derivative—Leibniz's notation	$(3x^3)' = 9x^2$
y''	second derivative	derivative of derivative	$(3x^3)'' = 18x$
$y^{(n)}$	nth derivative	n times derivation	$(3x^3)^{(3)} = 18$
$\dfrac{dy}{dx}$	derivative	derivative—Lagrange's notation	$d(3x^3)/dx = 9x^2$
$\dfrac{d^2y}{dx^2}$	second derivative	derivative of derivative	$d^2(3x^3)/dx^2 = 18x$
$\dfrac{d^ny}{dx^n}$	nth derivative	n times derivation	
$\dot{y}$	time derivative	derivative by time— Newton notation	
$\ddot{y}$	time second derivative	derivative of derivative	
$\dfrac{\partial f(x, y)}{\partial x}$	partial derivative		$\partial(x^2 + y^2)/\partial x = 2x$

(Continued)

Symbols	How to Read It	How to Use It	Examples
$\int$	integral	opposite to derivation	
$\iint$	double integral	integration of function of two variables	
$\iiint$	triple integral	integration of function of three variables	
$\oint$	closed contour/line integral		
$\oiint$	closed surface integral		
$\oiiint$	closed volume integral		
$[a,b]$	closed interval	$[a,b] = \{x \mid a \leq x \leq b\}$	
(a,b)	open interval	$(a,b) = \{x \mid a < x < b\}$	
i	imaginary unit	$i \equiv \sqrt{-1}$	$z = 3 + 2i$
z^*	complex conjugate	$z = a + bi \rightarrow z^* = a - bi$	$z^* = 3 + 2i$
$\overline{z}$	complex conjugate	$z = a + bi \rightarrow \overline{z} = a - bi$	$z = 3 + 2i$
∇	nabla/del	gradient/divergence operator	$\nabla f(x,y,z)$
$x^* y$	convolution	$y(t) = x(t) * h(t)$	
$\mathcal{L}$	Laplace transform	$F(s) = \mathcal{L}\{f(t)\}$	
$\mathcal{F}$	Fourier transform	$X(\omega) = \mathcal{F}\{f(t)\}$	
δ	delta function		

1.9 Numeral Symbols

Name	European	Roman	Hindu Arabic	Hebrew
zero	0		٠	
one	1	I	١	א
two	2	II	٢	ב
three	3	III	٣	ג
four	4	IV	٤	ד
five	5	V	٥	ה
six	6	VI	٦	ו
seven	7	VII	٧	ז
eight	8	VIII	٨	ח
nine	9	IX	٩	ט
ten	10	X	١٠	י
eleven	11	XI	١١	יא
twelve	12	XII	١٢	יב
thirteen	13	XIII	١٣	יג
fourteen	14	XIV	١٤	יד

(*Continued*)

(Continued)

Name	European	Roman	Hindu Arabic	Hebrew
fifteen	15	XV	١٥	טו
sixteen	16	XVI	١٦	טז
seventeen	17	XVII	١٧	יז
eighteen	18	XVIII	١٨	יח
nineteen	19	XIX	١٩	יט
twenty	20	XX	٢٠	כ
thirty	30	XXX	٣٠	ל
forty	40	XL	٤٠	מ
fifty	50	L	٥٠	נ
sixty	60	LX	٦٠	ס
seventy	70	LXX	٧٠	ע
eighty	80	LXXX	٨٠	פ
ninety	90	XC	٩٠	צ
one hundred	100	C	١٠٠	ק

1.10 Greek Alphabet Letters

Upper Case	Lower Case	Greek Letter Name
A	α	Alpha
B	β	Beta
Γ	γ	Gamma
Δ	δ	Delta
E	ε	Epsilon
Z	ζ	Zeta
H	η	Eta
Θ	θ	Theta
I	ι	Iota
K	κ	Kappa
Λ	λ	Lambda
M	μ	Mu
N	ν	Nu
Ξ	ξ	Xi
O	o	Omicron
Π	π	Pi
P	ρ	Rho
Σ	σ	Sigma
T	τ	Tau
Y	υ	Upsilon
Φ	φ	Phi

(Continued)

(Continued)

Upper Case	Lower Case	Greek Letter Name
X	χ	Chi
Ψ	ψ	Psi
Ω	ω	Omega

1.11 Roman Numerals

Number	Roman Numeral
1	I
2	II
3	III
4	IV
5	V
6	VI
7	VII
8	VIII
9	IX
10	X
11	XI
12	XII
13	XIII
14	XIV
15	XV
16	XVI
17	XVII
18	XVIII
19	XIX
20	XX
30	XXX
40	XL
50	L
60	LX
70	LXX
80	LXXX
90	XC
100	C
200	CC
300	CCC
400	CD

(*Continued*)

(Continued)

Number	Roman Numeral
500	D
600	DC
700	DCC
800	DCCC
900	CM
1000	M
5000	V
10,000	X
50,000	L
100,000	C
500,000	D
1,000,000	M

1.12 Prime Numbers

A prime number (or a prime) is a natural number >1 that has no positive divisors other than 1 and itself. Examples are as follows:

2, 3, 5, 7, 11, 13, 17, 19, 23, 29, 31, 37, 41, 43, 47, 53, 59, 61, 67, 71, 73, 79, 83, 89, 97, 101, 103, 107, 109, 113, 127, 131, 137, 139, 149, 151, 157, 163, 167, 173, 179, 181, 191, 193, 197, 199, 211, 223, 227, 229, 233, 239, 241, 251, 257, 263, 269, 271, 277, 281, . . .

1.13 Important Numbers in Science (Physical Constants)

- Avogadro constant (N_A) 6.02×10^{26} kmol^{-1}
- Boltzmann constant (k) 1.38×10^{-23} J K^{-1}
- Electron charge (e) 1.602×10^{-19} C
- Electron, charge/mass (e/m_e) 1.760×10^{11} C kg^{-1}
- Electron rest mass (m_e) 9.11×10^{-31} kg (0.511 MeV)
- Faraday constant (F) 9.65×10^4 C mol^{-1}
- Gas constant (R) 8.31×10^3 J K^{-1} kmol^{-1}
- Gas (ideal) normal volume (V_o) 22.4 m^3 kmol^{-1}
- Gravitational constant (G) 6.67×10^{-11} Nm^2 kg^{-2}
- Hydrogen atom (rest mass) (m_H) 1.673×10^{-27} kg (938.8 MeV)
- Neutron (rest mass) (m_n) 1.675×10^{-27} kg (939.6 MeV)
- Planck constant (h) 6.63×10^{-34} J s
- Proton (rest mass) (m_p) 1.673×10^{-27} kg (938.3 MeV)
- Speed of light (c) 3.00×10^8 m s^{-1}

1.14 Basic Conversion Formulas

When Converting from	Try Performing This Operation
Centimeters (cm) to feet (ft)	(cm) * 0.032808399 = (ft)
Centimeters (cm) to inches (in)	(cm) * 0.39370079 = (in)
Centimeters (cm) to meters (m)	(cm) * 0.01 = (m)
Centimeters (cm) to millimeters (mm)	(cm) * 10 = (mm)
Degrees (deg) to radians (rad)	(deg) * 0.01745329 = (rad)
Degrees Celsius (C) to degrees Fahrenheit (F)	[(C) * 1.8] + 32 = (F)
Degrees Fahrenheit (F) to degrees Celsius (C)	[(F) − 32)] * 0.555556 = (C)
Feet (ft) to centimeters (cm)	(ft) * 30.48 = (cm)
Feet (ft) to meters (m)	(ft) * 0.3048 = (m)
Feet (ft) to miles (mi)	(ft) * 0.000189393 = (mi)
Feet/minute (ft/min) to meters/second (m/s)	(ft/min) * 0.00508 = (m/s)
Feet/minute (ft/min) to miles/hour (mph)	(ft/min) * 0.01136363 = (mph)
Feet/second (ft/s) to kilometers/hour (kph)	(ft/s) * 1.09728 = (kph)
Feet/second (ft/s) to knots (kt)	(ft/s) * 0.5924838 = (kt)
Feet/second (ft/s) to meters/second (m/s)	(ft/s) * 0.3048 = (m/s)
Feet/second (ft/s) to miles/hour (mph)	(ft/s) * 0.681818 = (mph)
Inches (in) to centimeters (cm)	(in) * 2.54 = (cm)
Inches (in) to millimeters (mm)	(in) * 25.4 = (mm)
Kilometers (km) to meters (m)	(km) * 1000 = (m)
Kilometers (km) to miles (mi)	(km) * 0.62137119 = (mi)
Kilometers (km) to nautical miles (nmi)	(km) * 0.5399568 = (nmi)
Kilometers/hour (kph) to feet/second (ft/s)	(kph) * 0.91134 = (ft/s)
Kilometers/hour (kph) to knots (kt)	(kph) * 0.5399568 = (kt)
Kilometers/hour (kph) to meters/second (m/s)	(kph) * 0.277777 = (m/s)
Kilometers/hour (kph) to miles/hour (mph)	(kph) * 0.62137119 = (mph)
Knots (kt) to feet/second (ft/s)	(kt) * 1.6878099 = (ft/s)
Knots (kt) to kilometers/hour (kph)	(kt) * 1.852 = (kph)
Knots (kt) to meters/second (m/s)	(kt) * 0.514444 = (m/s)
Knots (kt) to miles/hour (mph)	(kt) * 1.1507794 = (mph)
Knots (kt) to nautical miles/hour (nmph)	Nothing—they are equivalent units
Meters (m) to centimeters (cm)	(m) * 100 = (cm)
Meters (m) to feet (ft)	(m) * 3.2808399 = (ft)
Meters (m) to kilometers (km)	(m) * 0.001 = (km)
Meters (m) to miles (mi)	(m) * 0.00062137119 = (mi)
Meters/second (m/s) to feet/minute (ft/min)	(m/s) * 196.85039 = (ft/min)
Meters/second (m/s) to feet/second (ft/s)	(m/s) * 3.2808399 = (ft/s)
Meters/second (m/s) to kilometers/hour (kph)	(m/s) * 3.6 = (kph)
Meters/second (m/s) to knots (kt)	(m/s) * 1.943846 = (kt)
Meters/second (m/s) to miles/hour (mph)	(m/s) * 2.2369363 = (mph)
Miles (mi) to feet (ft)	(mi) * 5280 = (ft)
Miles (mi) to kilometers (km)	(mi) * 1.609344 = (km)

(Continued)

(Continued)

When Converting from	Try Performing This Operation
Miles (mi) to meters (m)	(mi) * 1609.344 = (m)
Miles/hour (mph) to feet/minute (ft/min)	(mph) * 88 = (ft/min)
Miles/hour (mph) to feet/second (ft/s)	(mph) * 1.466666 = (ft/s)
Miles/hour (mph) to kilometers/hour (kph)	(mph) * 1.609344 = (kph)
Miles/hour (mph) to knots (kt)	(mph) * 0.86897624 = (kt)
Miles/hour (mph) to meters/second (m/s)	(mph) * 0.44704 = (m/s)
Millimeters (mm) to centimeters (cm)	(mm) * 0.1 = (cm)
Millimeters (mm) to inches (in)	(mm) * 0.039370078 = (in)
Nautical miles (nmi) to kilometers (km)	(nmi) * 1.852 = (km)
Nautical miles (nmi) to statute miles (mi)	(nmi) * 1.1507794 = (mi)
Nautical miles/hour (nmph) to knots (kt)	Nothing—they are equivalent units
Pounds/cubic foot (lb/ft^3) to kilograms/cubic meter (kg/m^3)	(lb/ft^3) * 16.018463 = (kg/m^3)
Radians (rad) to degrees (deg)	(rad) * 57.29577951 = (deg)
Statute miles (mi) to nautical miles (nmi)	(mi) * 0.86897624 = (nmi)

1.15   Basic Area Formulas

Square	side^2
Rectangle	length * width
Parallelogram	base * height
Triangle	base * height/2
Regular n-polygon	(1/4) * n * side^2 * cot(pi/n)
Trapezoid	height * (base1 + base2)/2
Circle	pi * radius^2
Ellipse	pi * radius1 * radius2
Cube (surface)	6 * side^2
Sphere (surface)	4 * pi * radius^2
Cylinder (surface of side)	perimeter of circle * height
	2 * pi * radius * height
Cylinder (whole surface)	areas of top and bottom circles + area of the side
	2(pi * radius^2) + 2 * pi * radius * height
Cone (surface)	pi * radius * side
Torus (surface)	pi^2 * (radius2^2 − radius1^2)

1.16   Basic Perimeter Formulas

Square	4 * side
Rectangle	2 * (length + width)
Parallelogram	2 * (side1 + side2)
Triangle	side1 + side2 + side3

(Continued)

	(Continued)
Regular n-polygon	n * side
Trapezoid	height * (base1 + base2)/2
Trapezoid	base1 + base2 + height * [csc(theta1) + csc(theta2)]
Circle	2 * pi * radius
Ellipse	4 * radius1 * $E(k,\text{pi}/2)$. $E(k,\text{pi}/2)$ is the complete elliptic integral of the second kind $k = (1/\text{radius1})$ * sqrt(radius1^2 − radius2^2)
Circumference or perimeter of a circle of radius r	$2\pi r$

1.17   Basic Volume Formulas

Cube	side^3
Rectangular prism	side1 * side2 * side3
Sphere	(4/3) * pi * radius^3
Ellipsoid	(4/3) * pi * radius1 * radius2 * radius3
Cylinder	pi * radius^2 * height
Cone	(1/3) * pi * radius^2 * height
Pyramid	(1/3) * (base area) * height
Torus	(1/4) * pi^2 * $(r1 + r2)$ * $(r1 − r2)^2$

2 Elementary Algebra

Elementary algebra encompasses some of the basic concepts of algebra, one of the main branches of mathematics. It is typically taught to secondary school students and builds on their understanding of arithmetic. Whereas arithmetic deals with specified numbers, algebra introduces quantities without fixed values known as variables. This use of variables entails a use of algebraic notation and an understanding of the general rules of the operators introduced in arithmetic. Unlike abstract algebra, elementary algebra is not concerned with algebraic structures outside the realm of real and complex numbers. The use of variables to denote quantities allows general relationships between quantities to be formally and concisely expressed, and thus enables solving a broader scope of problems. Most quantitative results in science and mathematics are expressed as algebraic equations.

- Sets of numbers
- Absolute value
- Basic properties of real numbers
- Logarithm
- Factorials
- Solving algebraic equations
- Intervals
- Complex numbers
- Euler's formula.

2.1 Sets of Numbers

Numbers are divided into two parts: real (IR) and complex numbers (C). Real numbers are divided into two types, rational numbers and irrational numbers.

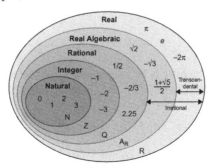

Mathematical Formulas for Industrial and Mechanical Engineering. DOI: http://dx.doi.org/10.1016/B978-0-12-420131-6.00002-6
© 2014 Elsevier Inc. All rights reserved.

Rational Numbers

1. Any number that can be expressed as the quotient of two integers (fraction). Example: 3/5.
2. Any number with a decimal that repeats or terminates. Example: 5.67676767... 10.345612.
3. Subsets of rational numbers:
   a. Integers: Rational numbers that contain no fractions or decimals $\{\ldots, -2, -1, 0, 1, 2, \ldots\}$.
   b. Whole numbers: All positive integers and the number 0 $\{0, 1, 2, 3, \ldots\}$.
   c. Natural numbers (counting numbers): All positive integers (not 0) $\{1, 2, 3, \ldots\}$.

Irrational Numbers

1. Any number that cannot be expressed as a quotient of two integers (fraction).
2. Any number with a decimal that does not repeat and does not terminate. Example: 4.34567129...
3. Most common example is π (3.14159265359...).

Complex Numbers

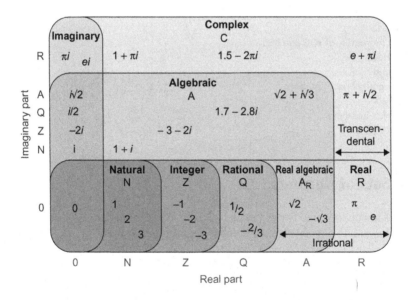

Many engineering problems can be treated and solved by using complex numbers. Many equations are not satisfied by any real numbers. Examples are

$$x^2 = -2 \quad \text{or} \quad x^2 - 2x + 40 = 0$$

We must introduce the concept of complex numbers.

Definition A complex number is an ordered pair $z = (x, y)$ of real numbers x and y. We call x the real art of z and y the imaginary part, and we write Re $z = x$, Im $z = y$. Example: $z = 5 + 2i$.

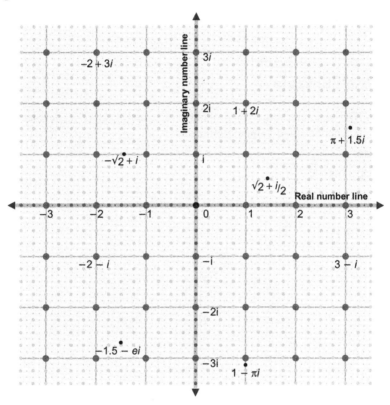

2.2 Fundamental Properties of Numbers

Linear inequalities: Linear inequalities in the real number system are the statements, such as $a > b$, $a < b$, $a \geq b$, $a \leq b$, where a & b are real numbers.

Definition

1. $a > b \Leftrightarrow a - b > 0$. Example: $5 > 3 \Leftrightarrow 5 - 3 = 2 > 0$
2. $a < b \Leftrightarrow b > a$. Example: $5 < 9 \Leftrightarrow 5 - 9 = -4 < 0$
3. $a \geq b \Leftrightarrow$ either $a > b$ or $a = b$
4. $a \leq b \Leftrightarrow$ either $a < b$ or $a = b$
5. $a < b < c \Leftrightarrow a < b$ & $b < c$. Example: $-3 < 5 < 11 \Leftrightarrow -3 < 5$ & $5 < 11$

Basic properties:

1. $a > b$, $b > c \Rightarrow a > c$. Example: $5 > 3$, $3 > 1 \Leftrightarrow 5 > 1$
   $a < b$, $b < c \Rightarrow a < c$. Example: $5 < 8$, $8 < 14 \Leftrightarrow 5 < 14$

2. $a > b \Rightarrow a \pm c > b \pm c$. Example: $5 > 3 \rightarrow 5 - 10 = -5 > 3 - 10 = -7$

$a < b \Rightarrow a \pm c < b \pm c$

3. $a > b, \ c > 0 \Rightarrow ac > bc, \ (a/c) > (b/c)$.   Example: $6 > 3 \rightarrow 6 \times 4 = 24 > 3 \times 4 = 12$,

$(6/2) = 3 > (3/2) = 1.5$

4. $a > b, c < 0 \Rightarrow ac < bc, \ (a/c) < (b/c)$. Example: $6 > 3 \rightarrow 6 \times -4 = -24 < 3 \times -4 = -12$,

$(6/-2) = -3 < (3/-2) = -1.5$

2.3 Absolute Value

Definition The absolute value (or modulus) of a real number x, denoted by the symbol $|x|$ is a nonnegative number defined as

$$|x| = \begin{cases} x, x \ge 0 \\ -x, x < 0 \end{cases}$$

For example, $|2| = 2, |0| = 0, |-2| = -(-2) = 2$.

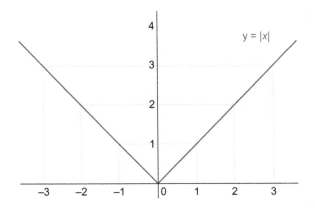

Geometrical interpretation: $|x|$ gives the distance of the point P, representing the number x on the real number line, from the origin.

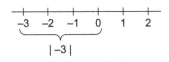

It is obvious that $|x| < a \Leftrightarrow -a < x < a$ & $|x| \le a \Leftrightarrow -a \le x \le a$.

Basic properties of modulus: The following properties of modulus are very useful in different types of problems, especially in mathematics.

1. $|x| \ge 0$

2. $x \le |x|, \ -x \le |x|$

3. $|x + y| \le |x| + |y|$

4. $|x - y| \ge |x| - |y|$

Example

Find the values of x, which satisfy the inequality $|2x - 3| \leq 4$.

Solution

$$|2x - 3| \leq 4 \Rightarrow -4 \leq 2x - 3 \leq 4 \Rightarrow -4 + 3 \leq 2x \leq 4 + 3 \Rightarrow -1 \leq 2x \leq 7 \Rightarrow -\frac{1}{2} \leq x \leq \frac{7}{2}$$

So, the values of x are in between $-1/2$ and $7/2$ inclusively.

2.4 Basic Properties of Real Numbers

1. *Closureness*: When two real numbers are added or multiplied together, we get again a real number. So, we say that the real number system is closed with respect to addition and multiplication. It is also closed with respect to subtraction. However, it is closed with respect to division, only when the divisor is non-zero.

In symbolic form, we write

   i. $a, b \in IR \Rightarrow a \pm b \in IR$ as well as $ab \in IR$

   ii. $a, b \in IR$ and $b \neq 0 \Rightarrow \dfrac{a}{b} \in IR$.

2. *Commutativity*: $a + b = b + a$ $\forall$ $a, b \in IR$. Example: $2 + 3 = 3 + 2$.
3. *Associativity*: $a + (b + c) = (a + b) + c$ $\forall$ $a, b, c \in IR$. Example: $2 + (3 + 4) = (2 + 3) + 4$.
4. *Distributivity*: $a \cdot (b + c) = a \cdot b + a \cdot c$ $\forall$ $a, b, c \in IR$. Example: $2 \cdot (3 + 4) = 2 \cdot 3 + 2 \cdot 4$.

   We say that multiplication distributes over addition. However, addition does not distribute over multiplication.

5. *Existence of identity elements*: $a + 0 = 0 + a$ and $a \cdot 1 = 1 \cdot a$ $\forall$ $a \in IR$. Example: $3 + 0 = 3$. $3 \cdot 1 = 3$.

   Here, the elements 0 & 1 are known as the *additive identity* and *multiplicative identity*, respectively.

6. Division by zero is undefined. Example: $5/0$ is undefined.

2.5 Laws of Exponents

For integers x and y:

1. $a^0 = 1$. Example: $7^0 = 1$
2. $a^x \cdot a^y = a^{x+y}$. Example: $4^5 \cdot 4^9 = 4^{14}$

3. $(a^x)^y = a^{xy}$. Example: $(4^5)^9 = 4^{45}$
4. $a^{-x} = 1/a^x$. Example: $4^{-5} = 1/4^5$
5. $a^{x/y} = \sqrt[y]{a^x}$. Example: $5^{3/7} = \sqrt[7]{5^3}$

2.6 Logarithm

1. $\log_a 1 = 0$, $\log_a a = 1$. Example: $\log_9 1 = 0, \log_{15} 15 = 1$
2. $\log_a x^m = m\log_a x$. Example: $\log_9 4^2 = 2\log_9 4$
3. $\log_a(xy) = \log_a x + \log_a y$. Example: $\log_9(7 \times 18) = \log_9 7 + \log_9 18$
4. $\log_a(x/y) = \log_a x - \log_a y$. Example: $\log_9(7/18) = \log_9 7 - \log_9 18$
5. $\log_a x = \log_a b \cdot \log_b x$. Example: $\log_9(7) = \log_9 15 \cdot \log_{15} 7$
6. $\log_a x = \dfrac{\log_b x}{\log_b a}$

Note: From (5), taking $x = a$, we get the formula:

$$\log_a b = \frac{1}{\log_b a}. \quad \text{Example: } \log_9 18 = \frac{1}{\log_{18} 9}$$

2.7 Factorials

The factorial of positive integer n is the product of all positive integers less than or equal to the integer n and is denoted by $n!$

$n! = 1 \times 2 \times 3 \times 4 \times \cdots \times n$. Example: $5! = 120$. By definition $0! = 1$.

Binomial expansion: For any value of n, whether positive, negative, integer, or noninteger, the value of the nth power of a binomial is given by

$$(x + y)^n = x^n + nx^{n-1}y + \frac{n(n-1)}{2!}x^{n-2}y^2 + \frac{n(n-1)(n-2)}{3!}x^{n-3}y^3 + \cdots + nxy^{n-1} + y^n$$

Example

$$(x + 3)^6 = x^6 + 6x^5(3) + 15x^4 3^2 + 20x^3 3^3 + 15x^2 3^4 + 6 \times 3^5 + 3^6$$

$$= x^6 + 18x^5 + 135x^4 3^2 + 540x^3 + 1215x^2 + 1458x + 729$$

$$(2x - 3)^5 = (2x)^5 + 5(2x)^4(-3) + 10(2x)^3(-3)^2 + 10(2x)^2(-3)^3 + 5(2x)(-3)^4 + (-3)^5$$

$$= 32x^5 - 240x^4 + 720x^3 - 1080x^2 + 810x - 243$$

$$(x^3 - 2/x)^4 = (x^3)^4 + 4(x^3)^3(-2/x) + 6(x^3)^2(-2/x)^2 + 4(x^3)(-2/x)^3 + (-2/x)^4$$

$$= x^{12} - 8x^8 + 24x^4 - 32 + 16x^{-4}$$

2.8   Factors and Expansions

$$(x + y)^2 = x^2 + 2xy + y^2$$

$$(x - y)^2 = x^2 - 2xy + y^2$$

$$(x + y)^3 = x^3 + 3x^2y + 3xy^2 + y^3$$

$$(x - y)^3 = x^3 - 3x^2y + 3xy^2 - y^3$$

$$(x + y)^4 = x^4 + 4x^3y + 6x^2y^2 + 4xy^3 + y^4$$

$$(x - y)^4 = x^4 - 4x^3y + 6x^2y^2 - 4xy^3 + y^4$$

$$(x^2 - y^2) = (x - y)(x + y)$$

$$(x^3 - y^3) = (x - y)(x^2 + xy + y^2)$$

$$(x^3 + y^3) = (x + y)(x^2 - xy + y^2)$$

$$(x^4 - y^4) = (x - y)(x + y)(x^2 + y^2)$$

$$(x^5 - y^5) = (x - y)(x^4 + x^3y + x^2y^2 + xy^3 + y^4)$$

2.9   Solving Algebraic Equations

Linear equation: $ax + b = c$. If $ax + b = c$ and $a \neq 0$, then the root is $x = (c - b)/a$.
Example: Solve $2x - 5 = 10 \rightarrow x = (10 + 5)/2 = 7.5$.
   Quadratic equation: $ax^2 + bx + c = 0$. If $ax^2 + bx + c = 0$, and $a \neq 0$, then roots are

$$x = \frac{-b \pm \sqrt{b^2 - 4ac}}{2a}$$

Example:

Solve $4x^2 + 5x + -9 = 0$, $\quad x = \dfrac{-5 \pm \sqrt{5^2 - 4(4)(-9)}}{2(4)} = \dfrac{-5 \pm 13}{8}$

   Cubic equation: $x^3 + bx^2 + cx + d = 0$. To solve $x^3 + bx^2 + cx + d = 0$, let $x = y - b/3$ then the reduced cubic is obtained as $y^3 + py + q = 0$, where

$p = c - (1/3)b^2$ and $q = d - (1/3)bc + (2/27)b^3$. The three roots of reduced cubic are

$$y_1 = \sqrt[3]{A} + \sqrt[3]{B}$$

$$y_2 = C\sqrt[3]{A} + C^2\sqrt[3]{B}$$

$$y_3 = C^2\sqrt[3]{A} + C\sqrt[3]{B}$$

where

$$A = -\frac{1}{2}q + \sqrt{(1/27)p^3 + (1/4)q^2}$$

$$B = -\frac{1}{2}q - \sqrt{(1/27)p^3 + (1/4)q^2}$$

$$C = \frac{-1 + i\sqrt{3}}{2}, \quad i^2 = -1$$

Then the solutions in terms of x:

$$x_1 = y_1 - (1/3)b$$

$$x_2 = y_2 - (1/3)b$$

$$x_3 = y_3 - (1/3)b$$

Quartic equation: $x^4 + ax^3 + bx^2 + cx + d = 0$

Let y_1 be a real root of the cubic equation: $y^3 - by^2 + (ac - 4d)y + (4bd - c^2 - a^2d) = 0$, then the solutions of the quartic equation are the roots of $z^2 + 0.5\left(a \pm \sqrt{a^2 - 4b + 4y_1}\right)z + 0.5\left(y_1 \pm \sqrt{y_1^2 - 4d}\right)$.

2.10 Intervals

Definition Given any two real numbers a and b, the set of all real numbers in between a and b is called an *interval*. Geometrically, an interval is a part of the real number line.

There are three types of interval:

i. Open interval: $(a, b) = \{x : a < x < b\}$
ii. Closed interval: $[a, b] = \{x : a \le x \le b\}$
iii. Left-open interval: $(a, b] = \{x : a < x \le b\}$
iv. Right-open interval: $[a, b) = \{x : a \le x < b\}$.

The last two types of interval are known as *semi-open* or *semi-closed* intervals.

Example

Express the following sets in terms of intervals.

a. $\{x \in R: -3 \leq x < 5\}$
b. $\{x \in R: |x| < 3\}$
c. $\{x \in R: |x - 1| \leq 2\}$

Solution

a. $[-3, 5)$
b. $|x| < 3 \Rightarrow -3 < x < 3 \Rightarrow (-3, 3)$
c. $|x - 1| \leq 2 \Rightarrow -2 \leq x - 1 \leq 2 \Rightarrow -2 + 1 \leq x \leq 2 + 1 \Rightarrow -1 \leq x \leq 3 \Rightarrow [-1, 3].$

Example

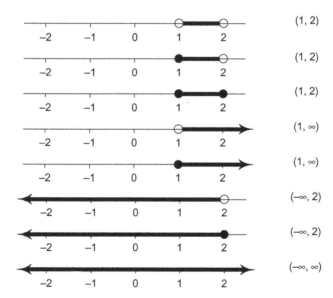

2.11   Complex Numbers

In algebra, we discovered that many equations are not satisfied by any real numbers. Examples are

$$x^2 = -2 \text{ or } x^2 - 2x + 40 = 0$$

We must introduce the concept of complex numbers.

Definition A complex number is an ordered pair $z = (x, y)$ of real numbers x and y. We call x the real part of z and y the imaginary part, and we write $\text{Re}\, z = x$, $\text{Im}\, z = y$.

Example Let $z = 5 - 2i$, $w = -2 + i$ and $u = 7i$.
  Then

$$\text{Re}(z) = 5 \qquad \text{Im}(z) = -2$$
$$\text{Re}(w) = -2 \quad \text{Im}(w) = 1$$
$$\text{Re}(u) = 0 \qquad \text{Im}(u) = 7$$

Two complex numbers are equal, where $z_1 = (x_1, y_1)$ and $z_2 = (x_2, y_2)$:

$$z_1 = z_2 \text{ if and only if } x_1 = x_2 \text{ and } y_1 = y_2$$

Addition and subtraction of complex numbers: We define for two complex numbers, the sum and difference of $z_1 = (x_1, y_1)$ and $z_2 = (x_2, y_2)$:

$$z_1 + z_2 = (x_1 + x_2, y_1 + y_2) \text{ and } z_1 - z_2 = (x_1 - x_2, y_1 - y_2).$$

Multiplication of two complex numbers is defined as follows:

$$z_1 z_2 = (x_1 x_2 - y_1 y_2, x_1 y_2 + x_2 y_1)$$

Example Let $z_1 = (3, 4)$ and $z_2 = (5, -6)$ then

$$z_1 + z_2 = (3 + 5, 4 + (-6)) = (8, -2)$$

and

$$z_1 - z_2 = (3 - 5, 4 - (-6)) = (-2, 10)$$

We need to represent complex numbers in a manner that will make addition and multiplication easier to do.
  Complex numbers represented as $z = x + iy$.
  A complex number whose imaginary part is 0 is of the form $(x, 0)$ and we have

$$(x_1, 0) + (x_2, 0) = (x_1 + x_2, 0) \text{ and } (x_1, 0) - (x_2, 0) = (x_1 - x_2, 0)$$

and

$$(x_1, 0) \cdot (x_2, 0) = (x_1 x_2, 0)$$

which looks like real addition, subtraction, and multiplication. So we identify $(x, 0)$ with the real number x and therefore we can consider the real numbers as a subset of the complex numbers.

We let the letter $i = (0, 1)$ and we call i a purely imaginary number. Now consider $i^2 = i \cdot i = (0, 1) \cdot (0, 1) = (-1, 0)$ and so we can consider the complex number $i^2 = -1 =$ the real number -1. We also get $yi = y \cdot (0, 1) = (0, y)$.

And so we have $(x, y) = (x, 0) + (0, y) = x + iy$.

Now we can write addition and multiplication as follows:

$$z_1 + z_2 = (x_1 + x_2, y_1 + y_2) = x_1 + x_2 + i(y_1 + y_2)$$

and

$$z_1 z_2 = (x_1 x_2 - y_1 y_2, x_1 y_2 + x_2 y_1) = x_1 x_2 - y_1 y_2 + i(x_1 y_2 + x_2 y_1)$$

Example Let $z_1 = (2, 3) = 2 + 3i$ and $z_2 = (5, -4) = 5 - 4i$, then

$$z_1 + z_2 = (2 + 3i) + (5 - 4i) = 7 - i$$

and

$$z_1 \cdot z_2 = (2 + 3i) \cdot (5 - 4i) = 10 + 15i - 8i - 12i^2 = 22 + 7i.$$

Example Solve the quadratic equation $x^2 + 4x + 5 = 0$. Using the quadratic formula, the solutions would be

$$x = \frac{-4 \pm \sqrt{4^2 - 4 \times 1 \times 5}}{2} = \frac{-4 \pm \sqrt{-4}}{2}$$

We notice a problem, however, since $\sqrt{-4}$ is not a real number. So the equation $x^2 + 4x + 5 = 0$ does not have any real roots.

However, suppose we introduced the symbol i to represent $\sqrt{-1}$. We could then find expressions for the solutions of the quadratic as

$$x = \frac{-4 \pm \sqrt{-4}}{2} = \frac{-4 \pm \sqrt{4 \times -1}}{2} = \frac{-4 \pm 2i}{2}$$

So the equation has two solutions: $x = -2 + i$ or $x = -2 - i$. These two solutions are called complex numbers.

2.12   The Complex Plane

The geometric representation of complex numbers is to represent the complex number (x, y) as the point (x, y).

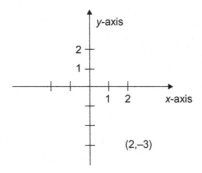

So the real number $(x, 0)$ is the point on the horizontal x-axis, the purely imaginary number $yi = (0, y)$ is on the vertical y-axis. For the complex number (x, y), x is the real part and y is the imaginary part. Example: Locate $2 - 3i$ on the graph above.

How do we divide complex numbers? Let's introduce the conjugate of a complex number then go to division.

Given the complex number $z = x + iy$, define the conjugate $\bar{z} = \overline{x + iy} = x - iy$. We can divide by using the following formula:

$$\frac{z_1}{z_2} = \frac{x_1 + iy_1}{x_2 + iy_2} = \frac{x_1 + iy_1}{x_2 + iy_2} \frac{x_2 - iy_2}{x_2 - iy_2} = \frac{x_1x_2 + y_1y_2 + i(x_2y_1 - x_1y_2)}{x_2^2 + y_2^2}$$

Example

$$\frac{2 + 3i}{3 - 4i} = \frac{(2 + 3i)(3 + 4i)}{(3 - 4i)(3 + 4i)} = \frac{6 + 12i^2 + 8i + 9i}{9 - 16i^2} = -\frac{6}{25} + i\frac{17}{25}$$

2.13   Complex Numbers in Polar Form

It is possible to express complex numbers in polar form. If the point $z = (x, y) = x + iy$ is represented by polar coordinates r, θ, then we can write $x = r \cos \theta$, $y = r \sin \theta$ and $z = r \cos \theta + ir \sin \theta = re^{i\theta}$. r is the modulus or absolute value of z, $|z| = r = \sqrt{x^2 + y^2}$, and θ is z, the argument of z, $\theta = \arctan(y/x)$. The values of r and θ determine z uniquely, but the converse is not true. The modulus r is determined uniquely by z, but θ is only determined up to a multiple of 2π. There are infinitely many values of θ which satisfy the equations $x = r \cos \theta, y = r \sin \theta$, but any two of them differ by some multiple of 2π. Each of these angles θ is called an argument of z, but, by convention, one of them is called the principal argument.

Definition If z is a non-zero complex number, then the unique real number θ, which satisfies

$$x = |z|\cos\theta, y = |z|\sin\theta, \quad -\pi < \theta \leq \pi$$

is called the principal argument of z, denoted by $\theta = \arg(z)$. *Note*: The distance from the origin to the point (x, y) is $|z|$, the modulus of z; the argument of z is the angle $\theta = \arctan(y/x)$. Geometrically, θ is the directed angle measured from the positive x-axis to the line segment from the origin to the point (x, y). When $z = 0$, the angle θ is undefined.

The polar form of a complex number allows one to multiply and divide complex numbers more easily than in the Cartesian form. For instance, if $z_1 = r_1 e^{i\theta_1}$ and $z_2 = r_2 e^{i\theta_2}$ then $z_1 z_2 = r_1 r_2 e^{i(\theta_1 + \theta_2)}$, $z_1/z_2 = (r_1/r_2)e^{i(\theta_1 - \theta_2)}$. These formulae follow directly from DeMoivre's formula.

Example For $z = 1 + i$, we get $r = \sqrt{1^2 + 1^2} = \sqrt{2}$ and $\theta = \arctan(y/x) = \arctan 1 = \pi/4$. The principal value of θ is $\pi/4$, but $9\pi/4$ would work also.

2.14 Multiplication and Division in Polar Form

Let $z_1 = r_1 \cos\theta_1 + ir_1 \sin\theta_1 = r_1(\cos\theta_1 + i\sin\theta_1)$ and $z_2 = r_2(\cos\theta_2 + i\sin\theta_2)$ then we have

$$z_1 z_2 = r_1 r_2(\cos(\theta_1 + \theta_2) + i\sin(\theta_1 + \theta_2)) \text{ and } \frac{z_1}{z_2} = \frac{r_1}{r_2}(\cos(\theta_1 - \theta_2) + i\sin(\theta_1 - \theta_2))$$

Example

$$z_1 = 1 + i = \sqrt{2}\left(\cos\frac{\pi}{4} + i\sin\frac{\pi}{4}\right) \text{ and } z_2 = \sqrt{3} - i = 2\left(\cos\frac{\pi}{6} + i\sin\frac{\pi}{6}\right)$$

Then

$$z_1 z_2 = \sqrt{2}\left(\cos\frac{\pi}{4} + i\sin\frac{\pi}{4}\right)2\left(\cos\frac{\pi}{6} + i\sin\frac{\pi}{6}\right) = 2\sqrt{2}\left(\cos\frac{5\pi}{12} + i\sin\frac{5\pi}{12}\right)$$

Since

$$\frac{\pi}{4} + \frac{\pi}{6} = \frac{10\pi}{24} = \frac{5\pi}{12}$$

and

$$\frac{z_1}{z_2} = \frac{\sqrt{2}(\cos(\pi/4) + i\sin(\pi/4))}{2(\cos(\pi/6) + i\sin(\pi/6))} = \frac{\sqrt{2}}{2}\left(\cos\frac{\pi}{12} + i\sin\frac{\pi}{12}\right)$$

We can use $z^2 = z \cdot z = r \cdot r(\cos(\theta + \theta) + i\sin(\theta + \theta)) = r^2(\cos 2\theta + i\sin 2\theta)$

2.15 DeMoivre's Theorem

$$z^n = r^n(\cos n\theta + i\sin n\theta)$$

where n is a positive integer.

Let $r = 1$ to get $(\cos\theta + i\sin\theta)^n = \cos n\theta + i\sin n\theta$.

Example Compute $(1+i)^6$

$$(1+i)^6 = \left(\sqrt{2}\left(\cos\frac{\pi}{4} + i\sin\frac{\pi}{4}\right)\right)^6$$

$$= \sqrt{2}^6\left(\cos 6\cdot\frac{\pi}{4} + i\sin 6\cdot\frac{\pi}{4}\right)$$

$$= 8\left(\cos\frac{3\pi}{2} + i\sin\frac{3\pi}{2}\right)$$

$$= -8i$$

2.16 Euler's Formula

$$e^{ix} = \cos x + i\sin x.$$

3 Linear Algebra

Linear algebra is the branch of mathematics concerning vector spaces, often finite or countable infinite dimensional, as well as linear mappings between such spaces. Such an investigation is initially motivated by a system of linear equations in several unknowns. Such equations are naturally represented using the formalism of matrices and vectors. Linear algebra is central to both pure and applied mathematics. For instance, abstract algebra arises by relaxing the axioms of a vector space, leading to a number of generalizations. Functional analysis studies the infinite-dimensional version of the theory of vector spaces. Combined with calculus, linear algebra facilitates the solution of linear systems of differential equations. Techniques from linear algebra are also used in analytic geometry, engineering, physics, natural sciences, computer science, computer animation, and the social sciences (particularly in economics). Because linear algebra is such a well-developed theory, nonlinear mathematical models are sometimes approximated by linear ones. Topics discussed in this chapter are as follows:

- Basic types of matrices
- Basic operations on matrices
- Determinants
- Sarrus rule
- Minors and cofactors
- Inverse matrix
- System of linear equations
- Cramer's rule.

3.1 Basic Definitions

Definition A matrix (plural form—matrices) is an arrangement of numbers in a rectangular form consisting of one or more rows and columns. Each number in the arrangement is called an *entry* or *element* of the matrix. A matrix is usually denoted by a capital letter and its elements are enclosed within square brackets [ ] or round brackets ( ) or double vertical bars $\| \ \|$.

If a_{ij} denotes the element in the ith row and jth column of a matrix A, then the matrix is written in the following form:

$$\begin{bmatrix} a_{11} & a_{12} & \dots & a_{1n} \\ a_{21} & a_{22} & \dots & a_{2n} \\ \vdots & \vdots & \vdots & \vdots \\ a_{m1} & a_{m2} & \dots & a_{mn} \end{bmatrix}$$

Mathematical Formulas for Industrial and Mechanical Engineering. DOI: http://dx.doi.org/10.1016/B978-0-12-420131-6.00003-8
© 2014 Elsevier Inc. All rights reserved.

Definition If a matrix has m rows and n columns, we call it a matrix of order m by n. The order of a matrix is also known as the *size* or *dimension* of the matrix.

For example, the matrices $A = \begin{bmatrix} 1 & 2 & 3 \\ -1 & 1 & 5 \end{bmatrix}$, $B = \begin{bmatrix} 2 & 1 \\ 3 & -1 \\ 1 & 5 \end{bmatrix}$, and $C = \begin{bmatrix} 1 & -3 \\ 2 & 5 \end{bmatrix}$ are of the orders 2×3, 3×2, and 2×2, respectively.

Definition Two matrices are said to be *equal* to each other if and only if they are of the same order and have the same corresponding elements.

3.2 Basic Types of Matrices

1. *Row matrix*: A matrix having a single row. Example: $\begin{bmatrix} 1 & -2 & 4 \end{bmatrix}$.

2. *Column matrix*: A matrix having a single column. Example: $\begin{bmatrix} -1 \\ 2 \\ 5 \end{bmatrix}$.

3. *Null matrix*: A matrix having all elements zero. Example: $\begin{pmatrix} 0 & 0 \\ 0 & 0 \end{pmatrix}$. A null matrix is also known as a *zero matrix*, and it is usually denoted by 0.

4. *Square matrix*: A matrix having equal number of rows and columns. Example: The matrix $\begin{pmatrix} 3 & -2 \\ -3 & 1 \end{pmatrix}$ is a square matrix of size 2×2.

5. *Diagonal matrix*: A square matrix, all of whose elements except those in the leading diagonal are zero. Example: $\begin{pmatrix} 2 & 0 & 0 \\ 0 & -3 & 0 \\ 0 & 0 & 5 \end{pmatrix}$.

6. *Scalar matrix*: A diagonal matrix having all the diagonal elements equal to each other. Example: $\begin{bmatrix} 3 & 0 & 0 \\ 0 & 3 & 0 \\ 0 & 0 & 3 \end{bmatrix}$.

7. *Unit matrix*: A diagonal matrix having all the diagonal elements equal to 1. Example: $\begin{bmatrix} 1 & 0 \\ 0 & 1 \end{bmatrix}$, $\begin{bmatrix} 1 & 0 & 0 \\ 0 & 1 & 0 \\ 0 & 0 & 1 \end{bmatrix}$, ... A unit matrix is also known as an identity matrix and is denoted by the capital letter I.

8. *Triangular matrix*: A square matrix, in which all the elements below (or above) the leading diagonal are zero. Example: $\begin{pmatrix} 3 & 1 & 4 \\ 0 & 2 & -1 \\ 0 & 0 & 4 \end{pmatrix}$ and $\begin{pmatrix} 1 & 0 & 0 \\ 2 & 3 & 0 \\ 4 & -1 & 5 \end{pmatrix}$ are upper triangular and lower triangular matrices, respectively.

9. *Symmetric matrix*: A square matrix $[a_{ij}]$ such that $a_{ij} = a_{ji} \; \forall ij$. Example: $\begin{bmatrix} 2 & 1 & -3 \\ 1 & 4 & 5 \\ -3 & 5 & 0 \end{bmatrix}$.

10. *Skew-symmetric matrix*: A square matrix $[a_{ij}]$ such that $a_{ij} = -a_{ji}$ $\forall$ ij.

Example: $\begin{bmatrix} 0 & 2 & 3 \\ -2 & 0 & 1 \\ -3 & -1 & 0 \end{bmatrix}$. Note that the elements in the leading diagonal of a skew-symmetric matrix are always zero.

3.3 Basic Operations on Matrices

We can obtain new matrices from the given ones by using the following operations:

1. *Addition and subtraction*: If $A = (a_{ij})$ and $B = (b_{ij})$ are $m \times n$ matrices, then their sum $A + B$ is defined as the new matrix: $(a_{ij} + b_{ij})$, where $1 \le i \le m$ and $1 \le j \le n$. The order of this sum is again $m \times n$. Similarly, $A - B$ is defined. Note that the sum or difference of two matrices is defined only when the matrices have the same size.
   Example:

   If $A = \begin{bmatrix} 3 & 4 & -2 \\ 2 & 0 & 1 \end{bmatrix}$, $B = \begin{bmatrix} 1 & 2 & 3 \\ 4 & -1 & 2 \end{bmatrix}$, and $C = \begin{bmatrix} -1 & 3 \\ 2 & 4 \end{bmatrix}$

   then $A + B$ and $A - B$ are defined, whereas $A + C$ and $A - C$ are not defined.

2. *Scalar multiplication*: If $A = (a_{ij})$ is a matrix and k is a scalar, then the scalar multiple of A by k, denoted by kA, is the matrix $B = (b_{ij})$ defined by $b_{ij} = ka_{ij}$. So, to multiply a given matrix A by a constant k means to multiply each element of A by k.
   Example:

   Let $A = \begin{pmatrix} 2 & 3 & -1 \\ 5 & 1 & 4 \end{pmatrix}$ and $k = 2$.

   Then, we get

   $$kA = 2\begin{pmatrix} 2 & 3 & -1 \\ 5 & 1 & 4 \end{pmatrix} = \begin{pmatrix} 4 & 6 & -2 \\ 10 & 2 & 8 \end{pmatrix}$$

3. *Multiplication*: If $A = (a_{ij})$ and $B = (b_{ij})$ are two matrices of orders $m \times p$ and $p \times n$, respectively, then the product AB is the new matrix $C = (c_{ij})$ of order $m \times n$ defined by the formula: $c_{ij} = a_{i1}b_{1j} + a_{i2}b_{2j} + a_{ip}b_{pj}$, where $1 \le i \le m$ and $1 \le j \le n$. In short, we write $c_{ij} = \sum_{k=1}^{p} a_{ik}b_{kj}$.
   Example:

   Let $A = \begin{pmatrix} 2 & 3 & -4 \\ 1 & 2 & 3 \end{pmatrix}$ and $B = \begin{pmatrix} 3 & 1 \\ -2 & 2 \\ 5 & -3 \end{pmatrix}$

   Then, we get

   $$AB = \begin{pmatrix} 2 \times 3 + 3 \times (-2) + (-4) \times 5 & 2 \times 1 + 3 \times 2 + (-4) \times (-3) \\ 1 \times 3 + 2 \times (-2) + 3 \times 5 & 1 \times 1 + 2 \times 2 + 3 \times (-3) \end{pmatrix} = \begin{pmatrix} -20 & 20 \\ 14 & -4 \end{pmatrix}$$

4. *Transposition*: The transpose of an $m \times n$ matrix $A = (a_{ij})$ is defined as the $n \times m$ matrix $A' = (a_{ji})$, where $1 \le i \le m$ and $1 \le j \le n$. It is also denoted as A^T.
   Example:

   If $A = \begin{bmatrix} 2 & -3 & 5 \\ 6 & 1 & 3 \end{bmatrix}$

then

$$A^T = \begin{bmatrix} 2 & 6 \\ -3 & 1 \\ 5 & 3 \end{bmatrix}$$

Remarks:

i. A is symmetric $\Leftrightarrow A^T = A$

ii. A is skew-symmetric $\Leftrightarrow A^T = -A$

3.4 Properties of Matrix Operations

Let A, B, and C be given matrices, then the basic properties of matrix addition, scalar multiplication, matrix multiplication, and matrix transposition are stated below without proof. These properties can be easily verified in examples.

1. *Properties of matrix addition and scalar multiplication*:
   i. $A + B = B + A$ (commutativity)
   ii. $(A + B) + C = A + (B + C)$ (associativity)
   iii. $A + O = O + A = A$, where O is the corresponding null matrix
   iv. $k(A + B) = kA + kB$, where k is a scalar (distributivity).
2. *Properties of matrix multiplication*:
   i. $AB \neq BA$, in general
   ii. $A(BC) = (AB)C$ (associativity)
   iii. $AI = IA = A$, where I is the corresponding identity matrix
   iv. $A(B + C) = AB + AC$ (distributivity).
3. *Properties of matrix transposition*:
   i. $(A^T)^T = A$
   ii. $(A + B)^T = A^T + B^T$
   iii. $(kA)^T = kA^T$
   iv. $(AB)^T = B^T A^T$.

Example 1

Given the matrices $\begin{bmatrix} 1 & 2 \\ -3 & 4 \end{bmatrix}$ and $\begin{bmatrix} 4 & 1 \\ 2 & 3 \end{bmatrix}$, verify that $AB \neq BA$.

Solution

We have, by multiplication:

$$AB = \begin{bmatrix} 1 & 2 \\ -3 & 4 \end{bmatrix} \cdot \begin{bmatrix} 4 & 1 \\ 2 & 3 \end{bmatrix} = \begin{bmatrix} 1 \times 4 + 2 \times 2 & 1 \times 1 + 2 \times 3 \\ (-3) \times 4 + 4 \times 2 & (-3) \times 1 + 4 \times 3 \end{bmatrix} = \begin{pmatrix} 8 & 7 \\ -4 & 9 \end{pmatrix}$$

$$BA = \begin{bmatrix} 4 & 1 \\ 2 & 3 \end{bmatrix} \cdot \begin{bmatrix} 1 & 2 \\ -3 & 4 \end{bmatrix} = \begin{bmatrix} 1 & 12 \\ -4 & 14 \end{bmatrix}$$

So, $AB \neq BA$.

Example 2

Given matrix is $A = \begin{pmatrix} 1 & 2 & -3 \\ -4 & 0 & 5 \end{pmatrix}$. Verify that $(A^T)^T = A$.

Solution

Here

$$A = \begin{pmatrix} 1 & 2 & -3 \\ -4 & 0 & 5 \end{pmatrix} \Rightarrow A^T = \begin{pmatrix} 1 & -4 \\ 2 & 0 \\ -3 & 5 \end{pmatrix} \Rightarrow (A^T)^T = \begin{pmatrix} 1 & 2 & -3 \\ -4 & 0 & 5 \end{pmatrix}$$

$\therefore (A^T)^T = A$. Hence, it is verified.

Example

Suppose that a company puts a problem to you:

Determine which of the three methods M_1, M_2, and M_3 of production it should use in producing three goods A, B, and C. The amount of each good produced by each method is shown below in matrix form:

$$\begin{array}{c} \\ M_1 \\ M_2 \\ M_3 \end{array} \begin{array}{ccc} A & B & C \\ \begin{pmatrix} 4 & 6 & 3 \\ 5 & 9 & 5 \\ 3 & 4 & 7 \end{pmatrix} \end{array}$$

The row matrix $(30 \quad 45 \quad 42)$ represents the profit per unit for the goods A, B, and C, respectively. Use matrix multiplication to decide which method maximizes the total profit.

Solution

Let $P = \begin{pmatrix} 4 & 6 & 3 \\ 5 & 9 & 5 \\ 3 & 4 & 7 \end{pmatrix}$ and $Q = (30 \quad 45 \quad 42)$. Then $Q' = \begin{pmatrix} 30 \\ 45 \\ 42 \end{pmatrix}$.

Now, the total profits from the three methods are given by

$$R = PQ' = \begin{pmatrix} 4 & 6 & 3 \\ 5 & 9 & 5 \\ 3 & 4 & 7 \end{pmatrix} \begin{pmatrix} 30 \\ 45 \\ 42 \end{pmatrix} = \begin{pmatrix} 120 + 270 + 126 \\ 150 + 405 + 210 \\ 90 + 180 + 294 \end{pmatrix} = \begin{pmatrix} 516 \\ 765 \\ 564 \end{pmatrix}$$

So, the profits obtained from the methods M_1, M_2, and M_3 are Rs. 516, Rs. 765, and Rs. 564, respectively. Consequently, the second method maximizes the profit.

3.5 Determinants

The determinant of a square matrix A is denoted by the symbol $|A|$ or det$A$. We can form determinants of $n \times n$ matrices. Such determinants are called $n \times n$ determinants.

Definition

1. If $A = [a_{11}]$ is a 1×1 matrix, then its determinant $|A|$ is equal to the number a_{11} itself.

2. If $A = \begin{bmatrix} a_{11} & a_{12} \\ a_{21} & a_{22} \end{bmatrix}$ is a 2×2 matrix, then the determinant is given by

$$|A| = \begin{vmatrix} a_{11} & a_{12} \\ a_{21} & a_{22} \end{vmatrix} = a_{11}a_{22} - a_{12}a_{21}$$

3. If $A = \begin{bmatrix} a_{11} & a_{12} & a_{13} \\ a_{21} & a_{22} & a_{23} \\ a_{31} & a_{32} & a_{33} \end{bmatrix}$ is a 3×3 matrix, then its determinant is given by

$$|A| = a_{11} \begin{vmatrix} a_{22} & a_{23} \\ a_{32} & a_{33} \end{vmatrix} - a_{12} \begin{vmatrix} a_{21} & a_{23} \\ a_{31} & a_{32} \end{vmatrix} + a_{13} \begin{vmatrix} a_{21} & a_{22} \\ a_{31} & a_{32} \end{vmatrix}$$

$$= a_{11}(a_{22}a_{33} - a_{23}a_{32}) - a_{12}(a_{21}a_{33} - a_{23}a_{31}) + a_{13}(a_{21}a_{32} - a_{22}a_{31})$$

Note that the determinant is expanded along the first row. Similarly, the determinant can be expanded along any other row or column carrying $(+)$ or $(-)$ sign according to the place occupied by the element in the following scheme:

$$\begin{vmatrix} + & - & + \\ - & + & - \\ + & - & + \end{vmatrix}$$

Example

1. $|2| = 2$

2. $\begin{vmatrix} 2 & 3 \\ 4 & -5 \end{vmatrix} = 2 \times (-5) - 3 \times 4 = -10 - 12 = -22$

3. $\begin{vmatrix} 1 & 3 & 5 \\ 2 & 1 & 3 \\ 3 & -4 & -6 \end{vmatrix} = 1 \begin{vmatrix} 1 & 3 \\ -4 & -6 \end{vmatrix} - 3 \begin{vmatrix} 2 & 3 \\ 3 & -6 \end{vmatrix} + 5 \begin{vmatrix} 2 & 1 \\ 3 & -4 \end{vmatrix}$, expanding along the first row

$$= 1(-6 + 12) - 3(-12 - 9) + 5(-8 - 3) = 6 + 63 - 55 = 14$$

3.6   Sarrus Rule

To find the value of a 3×3 determinant, the following rule, called the *Sarrus rule* may also be useful.

1. Consider the determinant: $\begin{vmatrix} a_{11} & a_{12} & a_{13} \\ a_{21} & a_{22} & a_{23} \\ a_{31} & a_{32} & a_{33} \end{vmatrix}$.

2. Write the three columns and then repeat the first two to have the fourth and fifth columns, respectively, as follows (vertical bars are to be omitted):

$$\begin{matrix} a_{11} & a_{12} & a_{13} & a_{11} & a_{12} \\ a_{21} & a_{22} & a_{23} & a_{21} & a_{22} \\ a_{31} & a_{32} & a_{33} & a_{31} & a_{32} \end{matrix}.$$

3. Write the product of the elements of each leading diagonal with positive sign. Also write the product of the elements of each secondary diagonal with negative sign.
4. The sum of the products obtained in step (3) gives the value of the determinant.

Example
Use Sarrus rule to find the value of the following determinant:

$$\begin{vmatrix} 1 & 3 & 5 \\ 2 & 1 & 3 \\ 3 & -4 & -6 \end{vmatrix}$$

Solution
The elements of the given determinant are arranged as follows:

$$\begin{matrix} 1 & 3 & 5 & 1 & 3 \\ 2 & 1 & 3 & 2 & 1 \\ 3 & -4 & -6 & 3 & -4 \end{matrix}$$

Now, according to Sarrus rule, the value of the determinant is

$$1 \times 1 \times (-6) + 3 \times 3 \times 3 + 5 \times 2 \times (-4) - 3 \times 1 \times 5 - (-4) \times 3 \times 1 - (-6)$$
$$\times 2 \times 3 = -6 + 27 - 40 - 15 + 12 + 36 = 75 - 61 = 14$$

3.7 Minors and Cofactors

Definition Let $D = \begin{vmatrix} a_{11} & a_{12} & a_{13} \\ a_{21} & a_{22} & a_{23} \\ a_{31} & a_{32} & a_{33} \end{vmatrix}$ be a 3×3 determinant.

Then the 2×2 determinant obtained by deleting the row and column, in which an element a_{ij} lies, is called the **minor** of the element and is denoted by M_{ij}.

Definition If M_{ij} is the minor of an element a_{ij} of a determinant, then the cofactor of the element, denoted by A_{ij} is defined as $A_{ij} = (-1)^{i+j} M_{ij}$.

Thus, the minor of a_{11} in D is $M_{11} = \begin{vmatrix} a_{22} & a_{23} \\ a_{32} & a_{33} \end{vmatrix}$ and the cofactor is $A_{11} = (-1)^{1+1} M_{11} = M_{11}$.

Remarks The value of the determinant D can be expressed in terms of the elements of any row or column of D.

Example

Write the cofactors of elements of second row of the determinant

$$D = \begin{vmatrix} 1 & 2 & 3 \\ -4 & 3 & 6 \\ 2 & -7 & 9 \end{vmatrix}$$

and hence find the value of the determinant.

Solution

Here

$$A_{21} = (-1)^{2+1} \begin{vmatrix} 2 & 3 \\ -7 & 9 \end{vmatrix} = -(18 + 21) = -39$$

$$A_{22} = (-1)^{2+2} \begin{vmatrix} 1 & 3 \\ 2 & 9 \end{vmatrix} = 9 - 6 = 3$$

$$A_{23} = (-1)^{2+3} \begin{vmatrix} 1 & 2 \\ 2 & -7 \end{vmatrix} = -(-7 - 4) = 11$$

So

$$A_{21} = -39, \quad A_{22} = 3, \quad \text{and} \quad A_{23} = 11$$

Now

$$D = a_{21}A_{21} + a_{22}A_{22} + a_{23}A_{23} = (-4)(-39) + 3(3) + 6(11) = 231$$

3.8   Properties of Determinants

Following properties may be helpful to find the values of determinants:

P1: If any two rows or columns of a determinant are identical, then its value equals zero.

Example

$$\begin{vmatrix} 2 & 3 & 1 \\ -3 & 4 & 2 \\ 2 & 3 & 1 \end{vmatrix} = 0 \text{ [The first and second rows are identical.]}$$

P2: If all the elements in any row or column of a determinant are zero, then the value of the determinant is also zero.

Example

$$\begin{vmatrix} 1 & -2 & 5 \\ 3 & 2 & 4 \\ 0 & 0 & 0 \end{vmatrix} = 0 \text{ [The elements of third row are all 0.]}$$

P3: The determinant of a unit matrix is equal to 1.

Example

$$\begin{vmatrix} 1 & 0 & 0 \\ 0 & 1 & 0 \\ 0 & 0 & 1 \end{vmatrix} = 1$$

P4: The determinant of a diagonal matrix is equal to the product of the diagonal elements.

Example

$$\begin{vmatrix} 2 & 0 & 0 \\ 0 & -3 & 0 \\ 0 & 0 & 4 \end{vmatrix} = 2 \times (-3) \times 4 = -24$$

P5: The determinant of a square matrix equals the determinant of its transpose.

Example

$$\begin{vmatrix} 1 & 2 & 3 \\ 3 & -1 & 2 \\ 2 & 0 & -3 \end{vmatrix} = \begin{vmatrix} 1 & 3 & 2 \\ 2 & -1 & 0 \\ 3 & 2 & -3 \end{vmatrix}$$

P6: If any two adjacent rows or columns of a determinant are interchanged, then the value of the determinant changes by sign.

Example

$$\begin{vmatrix} 1 & 2 & 3 \\ 3 & -1 & 2 \\ 2 & 0 & -3 \end{vmatrix} = - \begin{vmatrix} 3 & -1 & 2 \\ 1 & 2 & 3 \\ 2 & 0 & -3 \end{vmatrix}$$

P7: If each element of a row (or column) of a determinant is multiplied by a scalar k, then the value of the determinant is also multiplied by k.

Example

$$\begin{vmatrix} 3a & b & c \\ 3l & m & n \\ 3p & q & r \end{vmatrix} = 3 \begin{vmatrix} a & b & c \\ l & m & n \\ p & q & r \end{vmatrix}$$

P8: A determinant of the form $\begin{vmatrix} a+\alpha & b & c \\ l+\beta & m & n \\ p+\gamma & q & r \end{vmatrix}$ can be expressed in the form of the sum of two determinants:

$$\begin{vmatrix} a & b & c \\ l & m & n \\ p & q & r \end{vmatrix} + \begin{vmatrix} \alpha & b & c \\ \beta & m & n \\ \gamma & q & r \end{vmatrix}$$

P9: If a multiple of any row (or column) of a determinant is added to (or subtracted from) any other row (or column), then the value of the determinant remains unchanged.

Example

$$\begin{vmatrix} a & b & c \\ l & m & n \\ p & q & r \end{vmatrix} = \begin{vmatrix} a+kb & b & c \\ l+km & m & n \\ p+kq & q & r \end{vmatrix}$$

P10: The determinant of a product of two square matrices is equal to the product of their determinants, i.e., $|AB| = |A||B|$.

Further Examples

1. Evaluate the following determinants as indicated:

i. $\begin{vmatrix} 1 & 2 & 3 \\ 3 & -1 & 2 \\ 4 & 0 & -2 \end{vmatrix}$ (along the first row)

ii. $\begin{vmatrix} 1 & 2 & 3 \\ 3 & 7 & 4 \\ 2 & 3 & -2 \end{vmatrix}$ (along the first column)

Solution

i. $\begin{vmatrix} 1 & 2 & 3 \\ 3 & -1 & 2 \\ 4 & 0 & -2 \end{vmatrix} = 1\begin{vmatrix} -1 & 2 \\ 0 & -2 \end{vmatrix} - 2\begin{vmatrix} 3 & 2 \\ 4 & -2 \end{vmatrix} + 3\begin{vmatrix} 3 & -1 \\ 4 & 0 \end{vmatrix} = \cdots = 42$

ii. $\begin{vmatrix} 1 & 2 & 3 \\ 3 & 7 & 4 \\ 2 & 3 & -2 \end{vmatrix} = 1\begin{vmatrix} 7 & 4 \\ 3 & 5 \end{vmatrix} - 3\begin{vmatrix} 2 & 3 \\ 3 & 5 \end{vmatrix} + 2\begin{vmatrix} 2 & 3 \\ 7 & 4 \end{vmatrix} = \cdots = -6$

2. Evaluate the following determinant without expanding

$$\begin{vmatrix} 1 & 1 & 1 \\ a & b & c \\ b+c & c+a & a+b \end{vmatrix}$$

Solution

$$\begin{vmatrix} 1 & 1 & 1 \\ a & b & c \\ b+c & c+a & a+b \end{vmatrix} = \begin{vmatrix} 1 & 1 & 1 \\ a & b & c \\ a+b+c & a+b+c & a+b+c \end{vmatrix}; \quad R_3 \to R_3 + R_2$$

$$= (a+b+c)\begin{vmatrix} 1 & 1 & 1 \\ a & b & c \\ 1 & 1 & 1 \end{vmatrix}; \ (a+b+c) \text{ taken common from } R_3$$

$$= (a+b+c) \times 0 = 0$$

3. Prove that $\begin{vmatrix} 1 & 1 & 1 \\ a & b & c \\ a^2 & b^2 & c^2 \end{vmatrix} = (a-b)(b-c)(c-a)$

Solution

$$\begin{vmatrix} 1 & 1 & 1 \\ a & b & c \\ a^2 & b^2 & c^2 \end{vmatrix} = \begin{vmatrix} 1 & 0 & 0 \\ a & b-a & c-a \\ a^2 & b^2-a^2 & c^2-a^2 \end{vmatrix}; \quad C_2 \to C_2 - C_1 \text{ and } C_3 \to C_3 - C_1$$

$$= \begin{vmatrix} b-a & c-a \\ b^2-a^2 & c^2-a^2 \end{vmatrix}; \quad \text{expanding the determinant along } R_1$$

$$= (b-a)(c-a)\begin{vmatrix} 1 & 1 \\ b+a & c+a \end{vmatrix}; \quad \text{taking commons}$$

$$= (b-a)(c-a)(c-b)$$

$$= (a-b)(b-c)(c-a)$$

4. Solve the following equation:

$$\begin{vmatrix} 2 & -3 & 4 \\ -5 & 6 & -7 \\ 8 & -9 & x \end{vmatrix} = 0$$

Solution
Here, we have to find the value of x.
Now

$$\begin{vmatrix} 2 & -3 & 4 \\ -5 & 6 & -7 \\ 8 & -9 & x \end{vmatrix} = 0 \Rightarrow 8\begin{vmatrix} -3 & 4 \\ 6 & -7 \end{vmatrix} + 9\begin{vmatrix} 2 & 4 \\ -5 & -7 \end{vmatrix} + x\begin{vmatrix} 2 & -3 \\ -5 & 6 \end{vmatrix} = 0$$

$$\Rightarrow 8(21-24) + 9(-14+20) + x(12-15) = 0$$

$$\Rightarrow -24 + 54 - 3x = 0$$

$$\Rightarrow 3x = 30$$

$$\Rightarrow x = 10$$

3.9   Inverse Matrix

Definition A square matrix A is said to be *singular*, if its determinant $|A| = 0$ and nonsingular, if $|A| \neq 0$.

Definition The adjoint of a square matrix A, denoted as Adj.A, is defined as the transpose of the matrix obtained by replacing each element of A by its cofactor. So, Adj.$(A) = (A_{ji})$.

Example Let $A = \begin{pmatrix} 2 & 1 \\ 1 & 6 \end{pmatrix}$. Then the cofactors of its elements are: $A_{11} = 6, A_{12} = -1, A_{21} = -1,$ and $A_{22} = 2$. So, the matrix of cofactors is $\begin{pmatrix} 6 & -1 \\ -1 & 2 \end{pmatrix}$.

Definition If A and B are square matrices such that $AB = BA = I$, where I is the unit matrix of the same order, then B is called the *inverse (or reciprocal)* of A and is denoted by A^{-1}. Similarly, A is said to be the inverse of B.

Thus, $AA^{-1} = A^{-1}A = I$ and $BB^{-1} = B^{-1}B = I$.

Formula The inverse of a nonsingular matrix A is given by the formula:

$$A^{-1} = \frac{1}{A}(\mathrm{Adj}.A)$$

Note that no inverse of A exists, when $|A| = 0$.

Example

Find the inverse of the matrix $\begin{bmatrix} 0 & 1 & 2 \\ 1 & 2 & 3 \\ 3 & 1 & 1 \end{bmatrix}$.

Solution

Let $A = \begin{bmatrix} 0 & 1 & 2 \\ 1 & 2 & 3 \\ 3 & 1 & 1 \end{bmatrix}$. Then the determinant of A is $|A| = \begin{vmatrix} 0 & 1 & 2 \\ 1 & 2 & 3 \\ 3 & 1 & 1 \end{vmatrix} =$

$0(2 - 3) - 1(1 - 9) + 2(1 - 6) = -2$. Since $|A| \neq 0$. $\therefore A^{-1}$ exists.

Now, the cofactors of the elements of A are

$$A_{11} = \begin{vmatrix} 2 & 3 \\ 1 & 1 \end{vmatrix} = -1, \quad A_{12} = -\begin{vmatrix} 1 & 3 \\ 3 & 1 \end{vmatrix} = 8$$

$$A_{13} = \begin{vmatrix} 1 & 2 \\ 3 & 1 \end{vmatrix} = -5, \quad A_{21} = -\begin{vmatrix} 1 & 2 \\ 1 & 1 \end{vmatrix} = 1$$

$$A_{22} = \begin{vmatrix} 0 & 2 \\ 3 & 1 \end{vmatrix} = -6, \quad A_{23} = -\begin{vmatrix} 0 & 1 \\ 3 & 1 \end{vmatrix} = -3$$

$$A_{31} = \begin{vmatrix} 1 & 2 \\ 2 & 3 \end{vmatrix} = -1, \quad A_{32} = -\begin{vmatrix} 0 & 2 \\ 1 & 3 \end{vmatrix} = 2$$

$$A_{33} = \begin{vmatrix} 0 & 1 \\ 1 & 2 \end{vmatrix} = -1$$

$\therefore$ The matrix of cofactors is $\begin{bmatrix} -1 & 8 & 5 \\ 1 & -6 & 3 \\ 1 & 3 & -1 \end{bmatrix}$, hence

$$\text{Adj}(A) = \begin{bmatrix} -1 & 1 & -1 \\ 8 & -6 & 2 \\ -5 & 3 & -1 \end{bmatrix}.$$

Now

$$A^{-1} = \frac{1}{|A|} \text{Adj}(A) = \frac{1}{(-2)} \begin{bmatrix} -1 & 1 & -1 \\ 8 & -6 & 2 \\ -5 & 3 & -1 \end{bmatrix} = \begin{bmatrix} 1/2 & -1/2 & 1/2 \\ -4 & 3 & -1 \\ 5/2 & -3/2 & -1 \end{bmatrix}$$

$$\therefore A^{-1} = \begin{bmatrix} \dfrac{1}{2} & -\dfrac{1}{2} & \dfrac{1}{2} \\ -4 & 3 & -1 \\ \dfrac{5}{2} & -\dfrac{3}{2} & \dfrac{1}{2} \end{bmatrix}$$

3.10 System of Linear Equations

- An equation of the form $ax + by + c = 0$, where at least one of the real numbers a and b is not zero, is called a *linear equation* in x and y. Similarly, an equation of the form $ax + by + cz + d = 0$ is a linear equation in three variables x, y, and z.
- A system of equations can have exactly one solution, no solution, or an infinite number of solutions. A system with at least one solution is called a *consistent system*. If it has unique solution, it is said to be *consistent and independent*.

Example The system $\begin{cases} x + y = 5 \\ x - y = 3 \end{cases}$ has exactly one solution. The unique solution is $x = 4, y = 1$. So, it is a consistent and independent system.

- A system of equations having no solution is called an *inconsistent system*.

Example The system $\begin{cases} x + y = 5 \\ x + y = 7 \end{cases}$ has no solution, because no values of x and y satisfy both equations. Therefore, it is an inconsistent system.

- A system of equations is said to be consistent and dependent, if it has infinitely many solutions.

Example The system $\begin{cases} 2x + y = 7 \\ 4x + 2y = 14 \end{cases}$ is consistent, but dependent.

3.11   Methods of Solution

I. *Row-equivalent matrix method*: According to this method, to solve a system of linear equations in two variables, say, $\begin{cases} a_1x + b_1y = c_1 \\ a_2x + b_2y = c_2 \end{cases}$, we form the following matrix, called *augmented matrix*: $\begin{bmatrix} a_1 & b_1 & : & c_1 \\ a_2 & b_2 & : & c_2 \end{bmatrix}$.

Then, we use row operations to change this matrix into row-equivalent matrices. Some of the elementary row operations are as follows:

i. Interchange of any two rows, e.g., $R_1 \leftrightarrow R_2$.

ii. Multiplication of each element of a row by a non-zero number, e.g., $R_1 \rightarrow 3R_1$.

iii. Multiple of a row added to (or subtracted from) any other row, e.g., $R_2 \rightarrow R_2 + 2R_1$; $R_2 \rightarrow R_2 - 3R_1$.

In this way, row operations are performed until we get the following special form: $\begin{bmatrix} 1 & 0 & : & p \\ 0 & 1 & : & q \end{bmatrix}$. Then the solution of the system will be $x = p$, $y = q$.

Note: A system of three linear equations with three variables can be solved similarly.

Example 1

Find the solution, if any, of the system

$$\begin{cases} 2x - 3y = 4 \\ 5x + 4y = 1 \end{cases}$$

Solution
The augmented matrix is

$$\begin{bmatrix} 2 & -3 & : & 4 \\ 5 & 4 & : & 1 \end{bmatrix}$$

$$\sim \begin{bmatrix} 1 & -\dfrac{3}{2} & : & 2 \\ 5 & 4 & : & 1 \end{bmatrix}, \quad R_1 \rightarrow \frac{1}{2}R_1$$

$$\sim \begin{bmatrix} 1 & -\dfrac{3}{2} & : & 2 \\ 0 & \dfrac{23}{2} & : & -9 \end{bmatrix}, \quad R_2 \rightarrow R_2 - 5R_1$$

$$\sim \begin{bmatrix} 1 & -\dfrac{3}{2} & : & 2 \\ 0 & 1 & : & -\dfrac{18}{23} \end{bmatrix}, \quad R_2 \rightarrow \frac{2}{23}R_2$$

$$\sim \begin{bmatrix} 1 & 0 & : & \dfrac{19}{23} \\ 0 & 1 & : & -\dfrac{18}{23} \end{bmatrix}, \quad R_1 \rightarrow R_1 + \dfrac{3}{2}R_2$$

Thus the required solution is $x = 19/23$, $y = -18/23$.

Example 2

Solve the following system of equations:

$$\begin{cases} 2x + y + 6z = 3 \\ x - y + 4z = 1 \\ 3x + 2y - 2z = 2 \end{cases}$$

Solution

Starting with the corresponding augmented matrix and using elementary row operations, we get the following chain:

$$\begin{bmatrix} 2 & 1 & 6 & : & 3 \\ 1 & -1 & 4 & : & 1 \\ 3 & 2 & -2 & : & 2 \end{bmatrix} \sim \begin{bmatrix} 1 & -1 & 4 & : & 1 \\ 2 & 1 & 6 & : & 3 \\ 3 & 2 & -2 & : & 2 \end{bmatrix}, \quad R_1 \leftrightarrow R_2$$

$$\sim \begin{bmatrix} 1 & -1 & 4 & : & 1 \\ 0 & 3 & -2 & : & 1 \\ 0 & 5 & 14 & : & -1 \end{bmatrix}, \quad R_2 \rightarrow R_2 - 2R_1 \text{ and } R_3 \rightarrow R_3 - 3R_1$$

$$\sim \begin{bmatrix} 0 & -1 & 4 & : & 1 \\ 0 & 1 & -\dfrac{3}{2} & : & \dfrac{1}{3} \\ 0 & 5 & -14 & : & -1 \end{bmatrix}, \quad R_2 \rightarrow \dfrac{1}{3}R_2$$

$$\sim \begin{bmatrix} 1 & 0 & \dfrac{10}{3} & : & \dfrac{4}{3} \\ 0 & 1 & -\dfrac{2}{3} & : & \dfrac{1}{3} \\ 0 & 0 & -\dfrac{32}{3} & : & -\dfrac{8}{3} \end{bmatrix}, \quad R_1 \rightarrow R_1 + R_2 \text{ and } R_3 \rightarrow R_3 - 5R_2$$

$$\sim \begin{bmatrix} 1 & 0 & \frac{10}{3} & : & \frac{4}{3} \\ 0 & 1 & -\frac{2}{3} & : & \frac{1}{3} \\ 0 & 0 & 1 & : & \frac{1}{4} \end{bmatrix}, \quad R_3 \to -\frac{3}{32}R_3$$

$$\sim \begin{bmatrix} 1 & 0 & 0 & : & \frac{1}{2} \\ 0 & 1 & 0 & : & \frac{1}{2} \\ 0 & 0 & 1 & : & \frac{1}{4} \end{bmatrix}, \quad R_1 \to R_1 - \frac{10}{3}R_3 \text{ and } R_2 \to R_2 + \frac{2}{3}R_3$$

$\therefore$ The required solution is $x = \frac{1}{2}$, $y = \frac{1}{2}$, $z = \frac{1}{4}$.

3.12   Inverse Matrix Method

Consider the following system of linear equations:

$$\begin{cases} a_1 x + b_1 y = c_1 \\ a_2 x + b_2 y = c_2 \end{cases}.$$

This system can also be written in the form $\begin{bmatrix} a_1 & b_1 \\ a_2 & b_2 \end{bmatrix} \begin{bmatrix} x \\ y \end{bmatrix} = \begin{bmatrix} c_1 \\ c_2 \end{bmatrix}$.

So, we have $AX = B$, where $A = \begin{bmatrix} a_1 & b_1 \\ a_2 & b_2 \end{bmatrix}$, $X = \begin{bmatrix} x \\ y \end{bmatrix}$, and $B = \begin{bmatrix} c_1 \\ c_2 \end{bmatrix}$.
Now

$$\begin{aligned} AX &= B \\ &\Rightarrow A^{-1}(AX) = A^{-1}B \\ &\Rightarrow (A^{-1}A)X = A^{-1}B \\ &\Rightarrow IX = A^{-1}B \\ &\Rightarrow X = A^{-1}B \end{aligned}$$

For a system of three linear equations:

$$\begin{cases} a_1 x + b_1 y + c_1 z = d_1 \\ a_2 x + b_2 y + c_2 z = d_2 \\ a_3 x + b_3 y + c_3 z = d_3 \end{cases}$$

We have

$$
A = \begin{bmatrix} a_1 & b_1 & c_1 \\ a_2 & b_2 & c_2 \\ a_3 & b_3 & c_3 \end{bmatrix}, \quad X = \begin{bmatrix} x \\ y \\ z \end{bmatrix}, \quad \text{and } B = \begin{bmatrix} d_1 \\ d_2 \\ d_3 \end{bmatrix}
$$

So, the relation $X = A^{-1}B$ can be used to determine the values of x, y, and z that satisfy the system.

Example 1

Solve the system

$$
\begin{cases} 4x + 3y = 13 \\ 3x + y = -4 \end{cases}
$$

Solution

The given system can be written in the form

$$AX = B$$

where

$$
A = \begin{bmatrix} 4 & 5 \\ 3 & 1 \end{bmatrix}, \quad X = \begin{bmatrix} x \\ y \end{bmatrix}, \quad \text{and } B = \begin{bmatrix} 13 \\ -4 \end{bmatrix}
$$

The solution is given by $X = A^{-1}B$.
Let's first find A^{-1}. Here

$$
\text{Adj.}A = \begin{bmatrix} 1 & -5 \\ -3 & 4 \end{bmatrix} \quad \text{and } |A| = 4 \times 1 - 3 \times 5 = -11
$$

$$
\therefore A^{-1} = \frac{1}{|A|} \text{Adj.}A = \frac{1}{-11} \begin{bmatrix} 1 & -5 \\ -3 & 4 \end{bmatrix}
$$

Now

$$
X = A^{-1}B = \frac{1}{-11} \begin{bmatrix} 1 & -5 \\ -3 & 4 \end{bmatrix} \begin{bmatrix} 13 \\ -4 \end{bmatrix} = -\frac{1}{11} \begin{bmatrix} 33 \\ -55 \end{bmatrix} = \begin{bmatrix} -3 \\ 5 \end{bmatrix}
$$

$$
\Rightarrow \begin{bmatrix} x \\ y \end{bmatrix} = \begin{bmatrix} -3 \\ 5 \end{bmatrix}
$$

Hence the required solution is $x = -3$ and $y = 5$.

Example 2

Find the solution, if any, of the following system:

$$\begin{cases} x + 3y - z = -3 \\ 3x - y + 2z = 1 \\ 2x - y + 2z = -1 \end{cases}$$

Solution

The given system can be written as

$$\begin{bmatrix} 1 & 3 & -1 \\ 3 & -1 & 2 \\ 2 & -1 & 2 \end{bmatrix} \begin{bmatrix} x \\ y \\ z \end{bmatrix} = \begin{bmatrix} -3 \\ 1 \\ -1 \end{bmatrix}$$

So, we have $AX = B$, where

$$A = \begin{bmatrix} 1 & 3 & -1 \\ 3 & -1 & 2 \\ 2 & -1 & 2 \end{bmatrix}, X = \begin{bmatrix} x \\ y \\ z \end{bmatrix}, \text{ and } B = \begin{bmatrix} -3 \\ 1 \\ -1 \end{bmatrix}$$

Let's find A^{-1} first. The cofactors of the elements of A are

$$A_{11} = \begin{vmatrix} -1 & 2 \\ -1 & 2 \end{vmatrix} = 0, \qquad A_{12} = -\begin{vmatrix} 3 & 2 \\ 2 & 2 \end{vmatrix} = -2, \qquad A_{13} = \begin{vmatrix} 3 & -1 \\ 2 & -1 \end{vmatrix} = -1$$

$$A_{21} = -\begin{vmatrix} 3 & -1 \\ -1 & 2 \end{vmatrix} = -5, \quad A_{22} = \begin{vmatrix} 1 & -1 \\ 2 & 2 \end{vmatrix} = 4, \qquad A_{23} = -\begin{vmatrix} -1 & 3 \\ 2 & -1 \end{vmatrix} = -7$$

$$A_{31} = \begin{vmatrix} 3 & -1 \\ -1 & 2 \end{vmatrix} = 5, \qquad A_{32} = -\begin{vmatrix} 1 & -1 \\ 3 & 2 \end{vmatrix} = -5, \quad A_{33} = \begin{vmatrix} 1 & 3 \\ 3 & -1 \end{vmatrix} = -10$$

$\therefore$ The matrix of cofactors is

$$\begin{bmatrix} 0 & -2 & -1 \\ -5 & 4 & 7 \\ 5 & -5 & -10 \end{bmatrix}$$

Again

$$|A| = a_{11}A_{11} + a_{12}A_{12} + a_{13}A_{13} = 1 \times 0 + 3 \times (-2) + (-1) \times (-1) = -5 \neq 0$$

$$\therefore A^{-1} = \frac{1}{|A|} \text{Adj}.A = \frac{1}{-5} \begin{bmatrix} 0 & -5 & 5 \\ -2 & 4 & -5 \\ -1 & 7 & -10 \end{bmatrix}$$

Now

$$X = A^{-1}B = -\frac{1}{5}\begin{bmatrix} 0 & -5 & 5 \\ -2 & 4 & -5 \\ -1 & 7 & -10 \end{bmatrix}\begin{bmatrix} -3 \\ 1 \\ -1 \end{bmatrix} \Rightarrow \begin{bmatrix} x \\ y \\ z \end{bmatrix} = \begin{bmatrix} 2 \\ -3 \\ -4 \end{bmatrix}$$

Hence the required solution is $x = 2$, $y = -3$, and $z = -4$.

3.13 Determinant Method (Cramer's Rule)

Let us consider the system of equations: $\begin{cases} a_1x + b_1y = c_1 \\ a_2x + b_2y = c_2 \end{cases}$.

Multiplying the first equation by b_2, second by b_1, and then subtracting the second from the first, we get $(a_1b_2 - a_2b_1)x = b_2c_1 - b_1c_2$.

$$\therefore x = \frac{b_2c_1 - b_1c_2}{a_1b_2 - a_2b_1} = \frac{\begin{vmatrix} c_1 & b_1 \\ c_2 & b_2 \end{vmatrix}}{\begin{vmatrix} a_1 & b_1 \\ a_2 & b_2 \end{vmatrix}}$$

Similarly, we obtain $y = \dfrac{\begin{vmatrix} a_1 & c_1 \\ a_2 & c_2 \end{vmatrix}}{\begin{vmatrix} a_1 & b_1 \\ a_2 & b_2 \end{vmatrix}}$, provided that $\begin{vmatrix} a_1 & b_1 \\ a_2 & b_2 \end{vmatrix} \neq 0$.

Alternatively, we can write the above-mentioned formulae as shown below:
If

$$D = \begin{vmatrix} a_1 & b_1 \\ a_2 & b_2 \end{vmatrix}, \quad D_1 = \begin{vmatrix} c_1 & b_1 \\ c_2 & b_2 \end{vmatrix}, \quad \text{and } D_2 = \begin{vmatrix} a_1 & c_1 \\ a_2 & c_2 \end{vmatrix}$$

then the values of x and y are given by the formulae:

$$x = \frac{D_1}{D} \quad \text{and} \quad y = \frac{D_2}{D}$$

In the same way, for a system of three linear equations:

$$\begin{cases} a_1x + b_1y + c_1z = d_1 \\ a_2x + b_2y + c_2z = d_2 \\ a_3x + b_3y + c_3z = d_3 \end{cases}$$

the values of x, y, and z are given by the formulae:

$$x = \frac{D_1}{D}, y = \frac{D_2}{D}, \quad \text{and } z = \frac{D_3}{D}$$

where

$$D = \begin{vmatrix} a_1 & b_1 & c_1 \\ a_2 & b_2 & c_2 \\ a_3 & b_3 & c_3 \end{vmatrix}, \; D_1 = \begin{vmatrix} d_1 & b_1 & c_1 \\ d_2 & b_2 & c_2 \\ d_3 & b_3 & c_3 \end{vmatrix}, \; D_2 = \begin{vmatrix} a_1 & d_1 & c_1 \\ a_2 & d_2 & c_2 \\ a_3 & d_3 & c_3 \end{vmatrix},$$

$$\text{and } D_3 = \begin{vmatrix} a_1 & b_1 & d_1 \\ a_2 & b_2 & d_2 \\ a_3 & b_3 & d_3 \end{vmatrix}$$

It should be noted here that $D \neq 0$.

Example 1

Use Cramer's rule to solve the equations: $3x + 4y = 14, \quad 5x + 6y = 22.$

Solution
We have

$$D = \begin{vmatrix} 3 & 4 \\ 5 & 6 \end{vmatrix} = 3 \times 6 - 5 \times 4 = -2 \neq 0$$

$$D_1 = \begin{vmatrix} 14 & 4 \\ 22 & 6 \end{vmatrix} = 14 \times 6 - 4 \times 22 = -4$$

$$D_2 = \begin{vmatrix} 3 & 14 \\ 5 & 22 \end{vmatrix} = 3 \times 22 - 5 \times 14 = -4$$

Now, by Cramer's rule, we get

$$x = \frac{D_1}{D} = \frac{-4}{-2} = 2, \quad y = \frac{D_2}{D} = \frac{-4}{-2} = 2.$$

$$\therefore x = 2 \text{ and } y = 2.$$

Example

Use the determinant method to solve the following system of linear equations:

$$\begin{cases} 2x + y + z = 3 \\ -x + 2y + 2z = 1 \\ x - y - 3z = -6 \end{cases}$$

Solution
First of all, we calculate the following determinants:

$$D = \begin{vmatrix} 2 & 1 & 1 \\ -1 & 2 & 2 \\ 1 & -1 & -3 \end{vmatrix} = 2\begin{vmatrix} 2 & 2 \\ -1 & -3 \end{vmatrix} - 1\begin{vmatrix} -1 & 2 \\ 1 & -3 \end{vmatrix} + 1\begin{vmatrix} -1 & 2 \\ 1 & -1 \end{vmatrix} = \cdots = -10$$

$$D_1 = \begin{vmatrix} 2 & 1 & 1 \\ 1 & 2 & 2 \\ -6 & -1 & -3 \end{vmatrix} = 3 \begin{vmatrix} 2 & 2 \\ -1 & -3 \end{vmatrix} - 1 \begin{vmatrix} 1 & 2 \\ -6 & -3 \end{vmatrix} + 1 \begin{vmatrix} 1 & 2 \\ -6 & -1 \end{vmatrix} = \cdots = -10$$

$$D_2 = \begin{vmatrix} 2 & 3 & 1 \\ -1 & 1 & 2 \\ 1 & -6 & -3 \end{vmatrix} = 2 \begin{vmatrix} 1 & 2 \\ -6 & -3 \end{vmatrix} - 3 \begin{vmatrix} -1 & 2 \\ 1 & -3 \end{vmatrix} - 1 \begin{vmatrix} -1 & 1 \\ 1 & -6 \end{vmatrix} = \cdots = 20$$

$$D_3 = \begin{vmatrix} 2 & 1 & 3 \\ -1 & 2 & 1 \\ 1 & -1 & -6 \end{vmatrix} = 2 \begin{vmatrix} 2 & 1 \\ -1 & -6 \end{vmatrix} - 1 \begin{vmatrix} -1 & 1 \\ 1 & -6 \end{vmatrix} + 3 \begin{vmatrix} -1 & 2 \\ 1 & -1 \end{vmatrix} = \cdots = -30$$

Now, by Cramer's rule, we have

$$x = \frac{D_1}{D} = \frac{-10}{-10} = 1, \quad y = \frac{D_2}{D} = \frac{20}{-10} = -2, \quad z = \frac{D_3}{D} = \frac{-30}{-10} = 3$$

Hence, the required solution is $x = 1$, $y = -2$, and $z = 3$.

4 Analytic Geometry and Trigonometry

Geometry is divided into two branches: analytic geometry and trigonometry. Trigonometry began as the computational component of geometry. For instance, one statement of plane geometry states that a triangle is determined by a side and two angles. In other words, given one side of a triangle and two angles in the triangle, then the other two sides and the remaining angle are determined. Trigonometry includes the methods for computing those other two sides. The remaining angle is easy to find since the sum of the three angles equals 180 degrees (usually written as 180°). Analytic geometry is a branch of algebra that is used to model geometric objects—points, (straight) lines, and circles being the most basic of these. In plane analytic geometry (two-dimensional), points are defined as ordered pairs of numbers, say, (x, y), while the straight lines are in turn defined as the sets of points that satisfy linear equations. Topics discussed in this chapter are as follows:

- Plane figures
- Solid figures
- Triangles
- Degrees or radians
- Table of natural trigonometric functions
- Trigonometry identities
- The inverse trigonometric functions
- Solutions of trigonometric equations
- Analytic geometry (in the plane, i.e., 2D)
- Vector

Mathematical Formulas for Industrial and Mechanical Engineering. DOI: http://dx.doi.org/10.1016/B978-0-12-420131-6.00004-X
© 2014 Elsevier Inc. All rights reserved.

4.1 Plane Figures—Perimeter (*P*), Circumference (*C*), and Area (*A*)

Rectangle	Parallelogram	Trapezoid	Triangle	Circle
$P = 2l + 2w$				$C = 2\pi r$
$A = lw$	$A = bh$	$A = (1/2)(a + b)h$	$A = (1/2)bh$	$A = \pi r^2$

4.2 Solid Figures—Surface Area (*S*) and Volume (*V*)

Rectangular Solid	Circular Cylinder	Circular Cone	Sphere
$S = 2wl + 2hl + 2wh$	$S = 2\pi r^2 + 2\pi rh$	$S = \pi r^2 + \pi r\sqrt{(r^2 + h^2)}$	$S = 4\pi r^2$
$V = lwh$	$V = \pi r^2 h$	$V = (1/3)\pi r^2 h$	$V = (4/3)\pi r^3$

4.3 Right Triangle

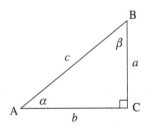

$c^2 = a^2 + b^2$ or $c = \sqrt{a^2 + b^2}$ (Pythagorean Theorem)
$\alpha + \beta = 90°$

$$\sin \alpha = \frac{a}{c} = \cos \beta$$

$$\cos \alpha = \frac{b}{c} = \sin \beta$$

$$\tan \alpha = \frac{a}{b} = \cot \beta$$

$$\cot \alpha = \frac{b}{a} = \tan \beta$$

$$\sec \alpha = \frac{c}{b} = \operatorname{cosec} \beta$$

$$\operatorname{cosec} \alpha = \frac{c}{a} = \sec \beta$$

4.4 Any Triangle

In any triangle with sides a, b, and c and corresponding opposite angles α, β, and γ:

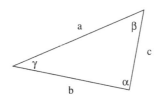

$\alpha + \beta + \gamma = 180°$
$a + b > c$
$b + c > a$
$a + c > a$

The law of sines is

$$\frac{a}{\sin \alpha} = \frac{b}{\sin \beta} = \frac{c}{\sin \gamma} \quad \text{or} \quad \frac{\sin \alpha}{a} = \frac{\sin \beta}{b} = \frac{\sin \gamma}{c}$$

The law of cosines is

$a^2 = b^2 + c^2 - 2cb \cos \alpha$
$b^2 = a^2 + c^2 - 2ca \cos \beta$
$c^2 = b^2 + a^2 - 2ab \cos \gamma$

The law of tangents is

$$\frac{a + b}{a - b} = \frac{\tan((\alpha + \beta)/2)}{\tan((\alpha - \beta)/2)}$$

$$\sin \frac{\alpha}{2} = \sqrt{\frac{(s-b)(s-c)}{bc}}, \text{ where } s = \frac{a+b+c}{2}$$

$$\cos \frac{\alpha}{2} = \sqrt{\frac{s(s-a)}{bc}}$$

$$\tan \frac{\alpha}{2} = \sqrt{\frac{(s-b)(s-c)}{s(s-a)}}$$

$$\text{Area} = \frac{bc \sin \alpha}{2}$$

Perimeter $= a + b + c$.

4.5   Degrees or Radians

Angles are measured in degrees or radians: $180° = \pi$ radians; 1 radian $= 180°/\pi$ degrees.

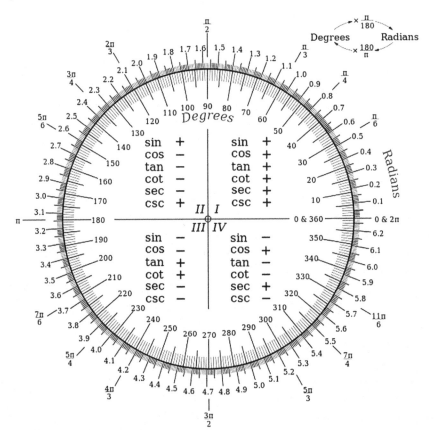

4.6 Table of Natural Trigonometric Functions

Radians	Degrees	Sin	Cos	Tan		
0.000	**00**	**0.0000**	**1.0000**	**0.0000**	**90**	**1.5707**
0.0175	01	0.0175	0.9998	0.0175	89	1.5533
0.0349	02	0.0349	0.9994	0.0349	88	1.5359
0.0524	03	0.0523	0.9986	0.0524	87	1.5184
0.0698	04	0.0698	0.9976	0.0699	86	1.5010
0.0873	05	0.0872	0.9962	0.0875	85	1.4835
0.1047	06	0.1045	0.9945	0.1051	84	1.4661
0.1222	07	0.1219	0.9925	0.1228	83	1.4486
0.1396	08	0.1392	0.9903	0.1405	82	1.4312
0.1571	09	0.1564	0.9877	0.1584	81	1.4137
0.1745	10	0.1736	0.9848	0.1763	80	1.3953
0.1920	11	0.1908	0.9816	0.1944	79	1.3788
0.2094	12	0.2079	0.9781	0.2126	78	1.3614
0.2269	13	0.2250	0.9744	0.2309	77	1.3439
0.2443	14	0.2419	0.9703	0.2493	76	1.3265
0.2618	**15**	**0.2588**	**0.9659**	**0.2679**	**75**	**1.3090**
0.2793	16	0.2756	0.9613	0.2867	74	1.2915
0.2967	17	0.2924	0.9563	0.3057	73	1.2741
0.3142	18	0.3090	0.9511	0.3249	72	1.2566
0.3316	19	0.3256	0.9455	0.3443	71	1.2392
0.3491	20	0.3420	0.9397	0.3640	70	1.2217
0.3665	21	0.3584	0.9336	0.3839	69	1.2043
0.3840	22	0.3746	0.9272	0.4040	68	1.1868
0.4014	23	0.3907	0.9205	0.4245	67	1.1694
0.4189	24	0.4067	0.9135	0.4452	66	1.1519
0.4363	25	0.4226	0.9063	0.4663	65	1.1345
0.4538	26	0.4384	0.8988	0.4877	64	1.1170
0.4712	27	0.4540	0.8910	0.5095	63	1.0996
0.4887	28	0.4695	0.8829	0.5317	62	1.0821
0.5061	29	0.4848	0.8746	0.5543	61	1.0647
0.5236	**30**	**0.5000**	**0.8660**	**0.5774**	**60**	**1.0472**
0.5411	31	0.5150	0.8572	0.6009	59	1.0297
0.5585	32	0.5299	0.8480	0.6249	58	1.0123
0.5760	33	0.5446	0.8387	0.6494	57	0.9948
0.5934	34	0.5592	0.8290	0.6745	56	0.9774
0.6109	35	0.5736	0.8192	0.7002	55	0.9599
0.6283	36	0.5878	0.8090	0.7265	54	0.9425
0.6458	37	0.6018	0.7986	0.7536	53	0.9250
0.6632	38	0.6157	0.7880	0.7813	52	0.9076
0.6807	39	0.6293	0.7771	0.8098	51	0.8901
0.6981	40	0.6428	0.7660	0.8391	50	0.8727
0.7156	41	0.6561	0.7547	0.8693	49	0.8552

(Continued)

(Continued)

Radians	Degrees	Sin	Cos	Tan		
0.7330	42	0.6691	0.7431	0.9004	48	0.8378
0.7505	43	0.6820	0.7314	0.9325	47	0.8203
0.7679	44	0.6947	0.7193	0.9657	46	0.8029
0.7854	**45**	**0.7071**	**0.7071**	**1.0000**	**45**	**0.7854**
		Cos	**Sin**	**Cot**	**Degrees**	**Radians**

P.S.: For an angle larger than 45° select such angle from right-hand side and obtain values in column corresponding to the function at bottom page. For the others functions like cot = 1/tan, csc = 1/sin, sec = 1/cos.

For an angle not found in the previous table, you can use the following rules:

$$\sin(-\theta) = -\sin\theta$$
$$\cos(-\theta) = \cos\theta$$
$$\tan(-\theta) = -\tan\theta$$

$$\sin\left(\frac{\pi}{2} - \theta\right) = \cos\theta \qquad \sin\left(\frac{\pi}{2} + \theta\right) = \cos\theta \qquad \sin\left(\frac{3\pi}{2} \pm \theta\right) = -\cos\theta$$

$$\cos\left(\frac{\pi}{2} - \theta\right) = \sin\theta \qquad \cos\left(\frac{\pi}{2} + \theta\right) = -\sin\theta \qquad \cos\left(\frac{3\pi}{2} \pm \theta\right) = \pm\sin\theta$$

$$\tan\left(\frac{\pi}{2} - \theta\right) = \cot\theta \qquad \tan\left(\frac{\pi}{2} + \theta\right) = -\cot\theta \qquad \tan\left(\frac{3\pi}{2} \pm \theta\right) = \pm\cot\theta$$

$$\sin(\pi - \theta) = \sin\theta \qquad \sin(\pi + \theta) = -\sin\theta \qquad \sin(2\pi \pm \theta) = \pm\sin\theta$$
$$\cos(\pi - \theta) = -\cos\theta \qquad \cos(\pi + \theta) = -\cos\theta \qquad \cos(2\pi \pm \theta) = \cos\theta$$
$$\tan(\pi - \theta) = -\tan\theta \qquad \tan(\pi + \theta) = \tan\theta \qquad \tan(2\pi \pm \theta) = \pm\tan\theta$$

4.7　Trigonometry Identities

Reciprocal Identities

$$\csc u = \frac{1}{\sin u} \qquad \sec u = \frac{1}{\cos u} \qquad \cot u = \frac{1}{\tan u}$$

$$\sin u = \frac{1}{\csc u} \qquad \cos u = \frac{1}{\sec u} \qquad \tan u = \frac{1}{\cot u}$$

Quotient Identities

$$\tan u = \frac{\sin u}{\cos u} \qquad \cot u = \frac{\cos u}{\sin u}$$

Cofunction Identities

$$\sin\left(\frac{\pi}{2} - u\right) = \cos u \quad \cos\left(\frac{\pi}{2} - u\right) = \sin u$$

$$\tan\left(\frac{\pi}{2} - u\right) = \cot u \quad \cot\left(\frac{\pi}{2} - u\right) = \tan u$$

$$\sec\left(\frac{\pi}{2} - u\right) = \csc u \quad \csc\left(\frac{\pi}{2} - u\right) = \sec u$$

Pythagorean Identities

1. $\sin^2\theta + \cos^2\theta = 1$
2. $\tan^2\theta + 1 = \sec^2\theta$
3. $\cot^2\theta + 1 = \csc^2\theta$

Sum and Difference of Angle Identities

1. $\sin(\alpha + \beta) = \sin\alpha\cos\beta + \cos\alpha\sin\beta$
2. $\sin(\alpha - \beta) = \sin\alpha\cos\beta - \cos\alpha\sin\beta$
3. $\cos(\alpha + \beta) = \cos\alpha\cos\beta - \sin\alpha\sin\beta$
4. $\cos(\alpha - \beta) = \cos\alpha\cos\beta + \sin\alpha\sin\beta$
5. $\tan(\alpha + \beta) = \dfrac{\tan\alpha + \tan\beta}{1 - \tan\alpha\tan\beta}$
6. $\tan(\alpha - \beta) = \dfrac{\tan\alpha - \tan\beta}{1 + \tan\alpha\tan\beta}$

Double Angle Identities

1. $\sin(2\theta) = 2\sin\theta\cos\theta$
2. $\cos(2\theta) = \cos^2\theta - \sin^2\theta$

$$= 2\cos^2 - 1$$

$$= 1 - 2\sin^2\theta$$

3. $\tan(2\theta) = \dfrac{2\tan\theta}{1 - \tan^2\theta}$

Half Angle Identities

1. $\sin\dfrac{\theta}{2} = \pm\sqrt{\dfrac{1 - \cos\theta}{2}}$

2. $\cos\dfrac{\theta}{2} = \pm\sqrt{\dfrac{1 + \cos\theta}{2}}$

3. $\tan\dfrac{\theta}{2} = \pm\sqrt{\dfrac{1 - \cos\theta}{1 + \cos\theta}}$

Product to Sum

$$\sin u \sin v = \frac{1}{2}(\cos(u-v) - \cos(u+v)) \quad \cos u \cos v = \frac{1}{2}(\cos(u-v) + \cos(u+v))$$

$$\sin u \cos v = \frac{1}{2}(\sin(u+v) + \sin(u-v)) \quad \cos u \sin v = \frac{1}{2}(\sin(u+v) - \sin(u-v))$$

Sum to Product

$$\sin u + \sin v = 2\sin\left(\frac{u+v}{2}\right)\cos\left(\frac{u-v}{2}\right) \quad \sin u - \sin v = 2\cos\left(\frac{u+v}{2}\right)\sin\left(\frac{u-v}{2}\right)$$

$$\cos u + \cos v = 2\cos\left(\frac{u+v}{2}\right)\cos\left(\frac{u-v}{2}\right) \quad \cos u - \cos v = -2\sin\left(\frac{u+v}{2}\right)\sin\left(\frac{u-v}{2}\right)$$

Power Reducing Identities

$$\sin^2 u = \frac{1 - \cos(2u)}{2}$$

$$\cos^2 u = \frac{1 + \cos(2u)}{2}$$

$$\tan^2 u = \frac{1 - \cos(2u)}{1 + \cos(2u)}$$

$$\sin^2 u = \frac{1 - \cos(2u)}{2}$$

$$\sin^3 u = \frac{3\sin u - \sin 3u}{4}$$

$$\cos^3 u = \frac{\cos 3u + 3\cos u}{4}$$

4.8   The Inverse Trigonometric Functions

Arc Sine

Let $f:[-\pi/2, \pi/2] \to [-1, 1]$ where $f(x) = \sin x$. Therefore, its inverse function is defined by $f^{-1}:[-1, 1] \to [-\pi/2, \pi/2]$, where $f^{-1}(x) = \sin^{-1} x$ $(1/\sin x)$ and is called the *arc sine function.*

Hence, if $y = \sin^{-1} x \Leftrightarrow \sin y = x$.

Arc Cosine

Let function g: $[0, \pi] \rightarrow [-1, 1]$, where $g(x) = \cos x$. Therefore, its inverse function is defined by g^{-1}: $[-1, 1] \rightarrow [0, \pi]$, where $g^{-1}(x) = \cos^{-1}x$ and is called the *arc cosine function*.

Also, $y = \cos^{-1}x \Leftrightarrow x = \cos y$.

Arc Tangent

Let h: $(-\pi/2, \pi/2) \rightarrow \Re$, where $h(x) = \tan x$. Therefore, its inverse function is defined by h^{-1}: $\Re \rightarrow (-\pi/2, \pi/2)$, where $h^{-1}(x) = \tan^{-1}x$ and is called the *arc tangent function*.

Hence, if $y = \tan^{-1}x \Leftrightarrow \tan y = x$.

4.9   Solutions of Trigonometric Equations

For $\tan \theta = k$, the general solution is $\theta = n\pi + \alpha$, $n \in Z$, $\alpha = \tan^{-1}k$.

For $\cos \theta = k$, where $|k| \leq 1$, the general solution is $\theta = 2n\pi \pm \alpha$, $n \in Z$, $\alpha = \cos^{-1}k$.

For $\sin \theta = k$, where $|k| \leq 1$, the general solution is $\theta = n\pi + (-1)^n\alpha$, $n \in Z$, $\alpha = \sin^{-1}k$.

4.10   Analytic Geometry (in the plane, i.e., 2D)

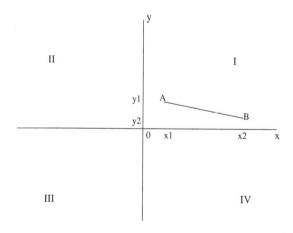

Given two points $A(x_1, y_1)$ and $B(x_2, y_2)$:

Distance formula: $D = \sqrt{(x_1 - x_2)^2 + (y_1 - y_2)^2}$

Midpoint formula: $M = \left(\dfrac{x_1 + x_2}{2} , \dfrac{y_1 + y_2}{2} \right)$

Slope formula: $m = \dfrac{y_2 - y_1}{x_2 - x_1}$

Slope and Angle of Inclination

	Slope	Angle of Inclination
Definition	Steepness of a line	Angle formed in relation to the x-axis
Notation	m	α
Computation	$\dfrac{Y_2 - Y_1}{X_2 - X_1}$	 *α is the angle formed between the lines
Relation	$m = \tan \alpha$	

Parallel lines have equal slopes.
Perpendicular lines have negative reciprocal slopes.
Equations for

Lines: $y = mx + b$; $y = m(x - x_1) + y_1$
Circles: $x^2 + y^2 = r^2$ (radius r, center the origin); $(x - h)^2 + (y - k)^2 = r^2$ (radius r, center (h, k))

A *tangent* to a circle is a line that touches the circle at only one point.
Theorem: A tangent to a circle is perpendicular to the radius to the point of tangency.

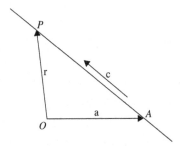

4.11 Vector

Vector equation of a straight line

$\vec{r} = \vec{a} + t\vec{c}$, t: scalar parameter

$\vec{a}$: position vector of a fixed point on the straight line

$\vec{c}$: direction vector

$\vec{r}$: position vector of any point on straight line

$\vec{r} = \vec{a} + t(\vec{b} - \vec{a})$

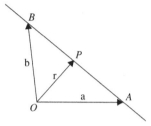

The length (magnitude) of the 2D vector $a = \langle a_1, a_2 \rangle$ is given by

$$|a| = \sqrt{a_1^2 + a_2^2}$$

The length (magnitude) of the 3D vector $a = \langle a_1, a_2, a_3 \rangle$ is given by

$$|a| = \sqrt{a_1^2 + a_2^2 + a_3^2}$$

Given a non-zero vector a, a *unit vector u* (vector of length one) in the same direction as the vector a can be constructed by multiplying a by the scalar quantity $1/|a|$, that is, forming

$$u = \frac{1}{|a|} \, a = \frac{a}{|a|}$$

The dot product of two vectors gives a *scalar* that is computed in the following manner:

In 2D, if $a = \langle a_1, a_2 \rangle$ and $b = \langle b_1, b_2 \rangle$, then dot product $= a \cdot b = a_1 b_1 + a_2 b_2$

In 3D, if $a = \langle a_1, a_2, a_3 \rangle$ and $b = \langle b_1, b_2, b_3 \rangle$, then dot product $= a \cdot b = a_1 b_1 + a_2 b_2 + a_3 b_3$

If the dot product $= 0$, the vectors are perpendicular

Angle between two vectors: Given two vectors a and b separated by an angle θ, $0 \le \theta \le \pi$.

Then

$$\cos \theta = \frac{a \cdot b}{|a| \, |b|}$$

Solving for θ gives

$$\theta = \arccos\left(\frac{a \cdot b}{|a| \ |b|}\right)$$

The cross product (it is defined only for vectors of length 3, i.e., 3D) of two vectors a and b gives a vector perpendicular to the plane of a and b that is computed in the following manner:

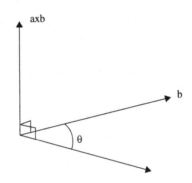

$$a \times b - \begin{vmatrix} i & j & k \\ a_1 & a_2 & a_3 \\ b_1 & b_2 & b_3 \end{vmatrix} - \begin{vmatrix} i & j & k \\ a_1 & a_2 & a_3 \\ b_1 & b_2 & b_3 \end{vmatrix} - \begin{vmatrix} i & j & k \\ a_1 & a_2 & a_3 \\ b_1 & b_2 & b_3 \end{vmatrix} + \begin{vmatrix} i & j & k \\ a_1 & a_2 & a_3 \\ b_1 & b_2 & b_3 \end{vmatrix}$$

$$- i \begin{vmatrix} a_2 & a_3 \\ b_2 & b_3 \end{vmatrix} - j \begin{vmatrix} a_1 & a_3 \\ b_1 & b_3 \end{vmatrix} + k \begin{vmatrix} a_1 & a_2 \\ b_1 & b_2 \end{vmatrix}$$

$$- i(a_2 b_3 - a_3 b_2) - j(a_1 b_3 - a_3 b_1) + k(a_1 b_2 - a_2 b_1)$$

5 Calculus

Calculus is the mathematical study of change, in the same way that geometry is the study of shape and algebra is the study of operations and their application to solving equations. It has two major branches: differential calculus (concerning rates of change and slopes of curves) and integral calculus (concerning accumulation of quantities and the areas under curves); these two branches are related to each other by the fundamental theorem of calculus. Both branches make use of the fundamental notions of convergence of infinite sequences and infinite series to a well-defined limit. Calculus has widespread uses in science, economics, and engineering and can solve many problems that algebra alone cannot. Topics discussed in this chapter are as follows:

- Functions and their graphs
- Limits of functions
- Definition and properties of the derivative
- Table of derivatives
- Applications of derivative
- Indefinite integral
- Integrals of rational function
- Integrals of irrational function
- Integrals of trigonometric functions
- Integrals of hyperbolic functions
- Integrals of exponential and logarithmic functions
- Reduction formulas using integration by part
- Definite integral
- Improper integral
- Continuity of a function
- Partial fractions
- Properties of trigonometric functions
- Sequences and series
- Convergence tests for series
- Taylor and Maclaurin series
- Continuous Fourier series
- Double integrals
- Triple integrals
- First-order differential equation
- Second-order differential equation
- Laplace transform
- Table of Laplace transforms

Mathematical Formulas for Industrial and Mechanical Engineering. DOI: http://dx.doi.org/10.1016/B978-0-12-420131-6.00005-1
© 2014 Elsevier Inc. All rights reserved.

5.1   Functions and Their Graphs

- A function $f(x)$ is a relation between a set of inputs (range) and a set of permissible outputs (domain) with the property that each input is related to exactly one output. Example of function: $f(x) = 5x$. Example of a non function: $f(x) = \pm\sqrt{x}$.
- Even function: $f(-x) = f(x)$. Example: $f(x) = x^2$.
- Odd function: $f(-x) = -f(x)$. Example: $f(x) = x^3$.
- Periodic function: $f(x + nT) = f(x)$. Example: $f(x) = \sin(x)$.
- Inverse function: $y = f(x)$ is any function, $x = g(y)$ or $y = f^{-1}(x)$ is its inverse function. Example: $f(x) = 2x + 3 \rightarrow f^{-1}(y) = (y - 3)/2$.
- Composite function: $y = f(u)$, $u = g(x)$, $y = f(g(x))$ is a composite function.
- Linear function: $y = ax + b$, $x \in R$, a slope of the line, b is the y-intercept. Example: $y = 2x - 10$.
- Quadratic function: $y = ax^2 + bx + c$, $x \in R$. Example: $y = x^2$.
- Cubic function: $y = ax^3 + bx^2 + cx + d$, $x \in R$. Example: $y = x^3$.
- Power function: $y = x^n$, $n \in N$.
- Square root function: $y = \sqrt{x}$, $x \in [0, \infty)$.
- Exponential functions: $y = a^x$, $a > 0$, $a \neq 1$, $y = e^x$, if $a = e$, $e = 2.71828182846\ldots$
- Logarithmic functions: $y = \log_a x$, $x \in (0, \infty)$, $a > 0$, $a \neq 1$, $y = \ln x$, if $a = e$, $x > 0$.
- Hyperbolic sine function: $y = \sinh x = \dfrac{e^x - e^{-x}}{2}$, $x \in R$.
- Hyperbolic cosine function: $y = \cosh x = \dfrac{e^x + e^{-x}}{2}$, $x \in R$.
- Hyperbolic tangent function: $y = \tanh x = \dfrac{\sin h\, x}{\cos h\, x} = \dfrac{e^x - e^{-x}}{e^x + e^{-x}}$, $x \in R$.
- Hyperbolic cotangent function: $y = \coth x = \dfrac{e^x + e^{-x}}{e^x - e^{-x}}$, $x \in R$, $x \neq 0$.
- Hyperbolic secant function: $y = \operatorname{sech} x = \dfrac{1}{\cosh x} = \dfrac{2}{e^x + e^{-x}}$, $x \in R$.
- Hyperbolic cosecant function: $y = \operatorname{csch} x = \dfrac{1}{\sinh x} = \dfrac{2}{e^x - e^{-x}}$, $x \in R$, $x \neq 0$.
- Inverse hyperbolic sine function: $y = \operatorname{arcsinh} x$, $x \in R$.
- Inverse hyperbolic cosine function: $y = \operatorname{arccosh} x$, $x \in [1, \infty)$.
- Inverse hyperbolic tangent function: $y = \operatorname{arctanh} x$, $x \in (-1, 1)$.
- Inverse hyperbolic cotangent function: $y = \operatorname{arccoth} x$, $x \in (-\infty, -1) \cup (1, \infty)$.
- Inverse hyperbolic secant function: $y = \operatorname{arcsech} x$, $x \in (0, 1]$.
- Inverse hyperbolic cosecant function: $y = \operatorname{arccsch} x$, $x \in R$, $x \neq 0$.

5.2   Limits of Functions

Functions: $f(x)$, $g(x)$
Variable: x
Real constants: a, k

- $\lim\limits_{x \to a}[f(x) + g(x)] = \lim\limits_{x \to a} f(x) + \lim\limits_{x \to a} g(x)$

- $\lim\limits_{x \to a}[f(x) - g(x)] = \lim\limits_{x \to a} f(x) - \lim\limits_{x \to a} g(x)$

- $\lim\limits_{x \to a}[f(x) \cdot g(x)] = \lim\limits_{x \to a} f(x) \cdot \lim\limits_{x \to a} g(x)$

- $\lim\limits_{x \to a} \dfrac{f(x)}{g(x)} = \dfrac{\lim\limits_{x \to a} f(x)}{\lim\limits_{x \to a} g(x)}, \quad$ if $\lim\limits_{x \to a} g(x) \neq 0$

- $\lim\limits_{x \to a}[kf(x)] = k \lim\limits_{x \to a} f(x)$

- $\lim\limits_{x \to a} f(g(x)) = f(\lim\limits_{x \to a} g(x))$

- $\lim\limits_{x \to a} f(x) = f(a)$, if the function $f(x)$ is continuous at $x = a$

- $\lim\limits_{x \to 0} \dfrac{\sin x}{x} = 1$

- $\lim\limits_{x \to 0} \dfrac{\tan x}{x} = 1$

- $\lim\limits_{x \to 0} \dfrac{\sin^{-1} x}{x} = 1$

- $\lim\limits_{x \to 0} \dfrac{\tan^{-1} x}{x} = 1$

- $\lim\limits_{x \to 0} \dfrac{\ln(1 + x)}{x} = 1$

- $\lim\limits_{x \to \infty} \left(1 + \dfrac{1}{x}\right)^x = e$

- $\lim\limits_{x \to \infty} \left(1 + \dfrac{k}{x}\right)^x = e^k$

- $\lim\limits_{x \to 0} a^x = 1$

5.3   Definition and Properties of the Derivative

Functions: f, g, y, u, v
Independent variable: x
Real constant: k
Angle: α

- $y'(x) = \lim\limits_{\Delta x \to 0} \dfrac{f(x + \Delta x) - f(x)}{\Delta x} = \lim\limits_{\Delta x \to 0} \dfrac{\Delta y}{\Delta x} = \dfrac{dy}{dx}$

- $\dfrac{dy}{dx} = \tan \alpha$

- $\dfrac{d(u+v)}{dx} = \dfrac{du}{dx} + \dfrac{dv}{dx}$

- $\dfrac{d(u-v)}{dx} = \dfrac{du}{dx} - \dfrac{dv}{dx}$

- $\dfrac{d(ku)}{dx} = k\dfrac{du}{dx}$

- Product rule:

$$\dfrac{d(u \cdot v)}{dx} = \dfrac{du}{dx} \cdot v + u \cdot \dfrac{dv}{dx}$$

- Quotient rule:

$$\dfrac{d}{dx}\left(\dfrac{u}{v}\right) = \dfrac{(du/dx \cdot v) - (u \cdot dv/dx)}{v^2}$$

- Chain rule:

$$y = f(g(x)), \quad u = g(x), \quad \dfrac{dy}{dx} = \dfrac{dy}{du} \cdot \dfrac{du}{dx}$$

- Derivative of inverse function:

$$\dfrac{dy}{dx} = \dfrac{1}{dx/dy}$$

   where $x(y)$ is the inverse function of $y(x)$.

- Reciprocal rule: $\dfrac{d}{dx}\left(\dfrac{1}{y}\right) = -\dfrac{dy/dx}{y^2}$

- Logarithmic differentiation:

   - $y = f(x),\ \ln y = \ln f(x),$

   - $\dfrac{dy}{dx} = f(x) \cdot \dfrac{d}{dx}[\ln f(x)].$

5.4 Table of Derivatives

Independent variable: x
Real constants: C, a, b, c
Natural number: n

- $\dfrac{d}{dx}(C) = 0$

- $\dfrac{d}{dx}(x) = 1$

- $\dfrac{d}{dx}(ax+b) = a$

- $\dfrac{d}{dx}(ax^2+bx+c) = ax+b$

- $\dfrac{d}{dx}(x^n) = nx^{n-1}$

- $\dfrac{d}{dx}(x^{-n}) = -\dfrac{n}{x^{n+1}}$

- $\dfrac{d}{dx}\left(\dfrac{1}{x}\right) = -\dfrac{1}{x^2}$

- $\dfrac{d}{dx}\left(\sqrt{x}\right) = \dfrac{1}{2\sqrt{x}}$

- $\dfrac{d}{dx}\left(\sqrt[n]{x}\right) = \dfrac{1}{n\sqrt[n]{x^{n-1}}}$

- $\dfrac{d}{dx}(\ln x) = \dfrac{1}{x}$

- $\dfrac{d}{dx}(\log_a x) = \dfrac{1}{x\ln a}, \quad a>0, \ a\neq 1$

- $\dfrac{d}{dx}(a^x) = a^x \ln a, \quad a>0, \ a\neq 1$

- $\dfrac{d}{dx}(e^x) = e^x$

- $\dfrac{d}{dx}(\sin x) = \cos x$

- $\dfrac{d}{dx}(\cos x) = -\sin x$

- $\dfrac{d}{dx}(\tan x) = \dfrac{1}{\cos^2 x} = \sec^2 x$

- $\dfrac{d}{dx}(\cot x) = -\dfrac{1}{\sin^2 x} = -\csc^2 x$

- $\dfrac{d}{dx}(\sec x) = \tan x \cdot \sec x$

- $\dfrac{d}{dx}(\csc x) = -\cot x \cdot \csc x$

- $\dfrac{d}{dx}(\arcsin x) = \dfrac{1}{\sqrt{1-x^2}}$

- $\dfrac{d}{dx}(\arccos x) = -\dfrac{1}{\sqrt{1 - x^2}}$

- $\dfrac{d}{dx}(\arctan x) = \dfrac{1}{1 + x^2}$

- $\dfrac{d}{dx}(\text{arccot } x) = -\dfrac{1}{1 + x^2}$

- $\dfrac{d}{dx}(\text{arcsec } x) = \dfrac{1}{|x|\sqrt{x^2 - 1}}$

- $\dfrac{d}{dx}(\text{arccsc } x) = -\dfrac{1}{|x|\sqrt{x^2 - 1}}$

- $\dfrac{d}{dx}(\sinh x) = \cosh x$

- $\dfrac{d}{dx}(\cosh x) = \sinh x$

- $\dfrac{d}{dx}(\tanh x) = \dfrac{1}{\cosh^2 x} = \text{sech}^2 x$

- $\dfrac{d}{dx}(\coth x) = -\dfrac{1}{\sinh^2 x} = -\text{csch}^2 x$

- $\dfrac{d}{dx}(\text{sech } x) = -\text{sech } x \cdot \tanh x$

- $\dfrac{d}{dx}(\text{csch } x) = -\text{csch } x \cdot \coth x$

- $\dfrac{d}{dx}(\text{arcsinh } x) = \dfrac{1}{\sqrt{x^2 + 1}}$

- $\dfrac{d}{dx}(\text{arccosh } x) = \dfrac{1}{\sqrt{x^2 - 1}}$

- $\dfrac{d}{dx}(\text{arctanh } x) = \dfrac{1}{1 - x^2}, \quad |x| < 1$

- $\dfrac{d}{dx}(\text{arccoth } x) = -\dfrac{1}{x^2 - 1}, \quad |x| > 1$

- $\dfrac{d}{dx}(u^v) = vu^{v-1} \cdot \dfrac{du}{dx} + u^v \ln u \cdot \dfrac{dv}{dx}$

5.5   Higher Order Derivatives

Functions: f, y, u, v
Independent variable: x
Natural number: n

- Second derivative:

$$f'' = (f')' = \left(\frac{dy}{dx}\right)' = \frac{d}{dx}\left(\frac{dy}{dx}\right) = \frac{d^2y}{dx^2}$$

- Higher order derivative:

$$f^{(n)} = \frac{d^n y}{dx^n} = y^{(n)} = (f^{(n-1)})'$$

- $(u+v)^{(n)} = u^{(n)} + v^{(n)}$

- $(u-v)^{(n)} = u^{(n)} - v^{(n)}$

- Leibnitz's formulas:

  - $(uv)'' = u''v + 2u'v' + uv''$

  - $(uv)''' = u'''v + 3u''v' + 3u'v'' + uv'''$

  - $(uv)^{(n)} = u^{(n)}v + nu^{(n-1)}v' + \dfrac{n(n-1)}{1.2}u^{(n-2)}v' + \cdots + uv^{(n)}$

- $(x^m)^{(n)} = \dfrac{m!}{(m-n)!}x^{m-n}$

- $(x^n)^{(n)} = n!$

- $(\log_a x)^{(n)} = \dfrac{(-1)^{n-1}(n-1)!}{x^n \ln a}$

- $(\ln x)^{(n)} = \dfrac{(-1)^{n-1}(n-1)!}{x^n}$

- $(a^x)^{(n)} = a^x \ln^n a$

- $(e^x)^{(n)} = e^x$

- $(a^{mx})^{(n)} = m^n a^{mx} \ln^n a$

- $(\sin x)^{(n)} = \sin\left(x + \dfrac{n\pi}{2}\right)$

- $(\cos x)^{(n)} = \cos\left(x + \dfrac{n\pi}{2}\right)$

5.6   Applications of Derivative

Functions: f, g, y
Position of an object: s
Velocity: v
Acceleration: w
Independent variable: x
Time: t
Natural number: n

* Velocity and acceleration
  $s = f(t)$ is the position of an object relative to a fixed coordinate system at a time t, $v = s' = f'(t)$ is the instantaneous velocity of the object, $w = v' = s'' = f''(t)$ is the instantaneous acceleration of the object.
* Tangent line: $y - y_0 = f'(x_0)(x - x_0)$.
* Normal line:

$$y - y_0 = -\frac{1}{f'(x_0)}(x - x_0)$$

* Increasing and decreasing functions:
  If $f'(x_0) > 0$, then $f(x)$ is increasing at x_0.
  If $f'(x_0) < 0$, then $f(x)$ is decreasing at x_0.
  If $f'(x_0)$ does not exist or is zero, then the test fails.
* Local extrema: A function $f(x)$ has a local maximum at x, if and only if there exists some interval containing x, such that $f(x_1) \geq f(x)$ for all x in the interval.
* A function $f(x)$ has a local minimum at x_2, if and only if there exists some interval containing x_2, such that $f(x_2) \leq f(x)$ for all x in the interval.
* Critical points: A critical point on $f(x)$ occurs at x_0, if and only if either $f(x_0)$ is zero or the derivative doesn't exist.
* First derivative test for local extrema:
  If $f(x)$ is increasing $(f'(x) > 0)$ for all x in some interval $(a, x_1]$ and $f(x)$ is decreasing $(f'(x) < 0)$ for all x in some interval $[x_1, b)$, then $f(x)$ has a local maximum at x_1.
* If $f(x)$ is decreasing $(f'(x) < 0)$ for all x in some interval $(a, x_2]$ and $f(x)$ is increasing $(f'(x) > 0)$ for all x in some interval $[x_2, b)$, then $f(x)$ has a local minimum at x_2.
* Second derivative test for local extrema:
  If $f'(x_1) = 0$ and $f''(x_1) < 0$, then $f(x)$ has a local maximum at x_1.
  If $f'(x_2) = 0$ and $f''(x_2) > 0$, then $f(x)$ has a local minimum at x_2.
* Concavity:
  $f(x)$ is concave upward at x_0 if and only if $f'(x)$ is increasing at x_0.
  $f(x)$ is concave downward at x_0 if and only if $f'(x)$ is decreasing at x_0.
* Second derivative test for concavity:
  If $f''(x_0) > 0$, then $f(x)$ is concave upward at x_0.
  If $f''(x_0) < 0$, then $f(x)$ is concave downward at x_0.
  If $f''(x)$ does not exist or is zero, then the test fails.
* Inflection points
  If $f'(x_3)$ exists and $f''(x)$ changes sign at $x = x_3$, then the point $(x_3, f(x_3))$ is an inflection point of the graph of $f(x)$. If $f''(x_3)$ exists at the inflection point, then $f''(x_3) = 0$.

- L'Hopital's rule:

$$\lim_{x \to c} \frac{f(x)}{g(x)} = \lim_{x \to c} \frac{f'(x)}{g'(x)}, \quad \text{if} \quad \lim_{x \to c} f(x) = \lim_{x \to c} g(x) = 0 \quad \text{or} \quad \infty.$$

5.7 Indefinite Integral

Functions: f, g, u, v
Independent variables: x, t, ξ
Indefinite integral of a function:

$$\int f(x)dx, \quad \int g(x)dx, \ldots$$

Derivative of a function:

$$y'(x), \quad f'(x), \quad F'(x), \ldots$$

Real constants: C, a, b, c, d, k
Natural numbers: m, n, i, j

- $\int f(x)dx = F(x) + C, \quad \text{if} \quad F'(x) = f(x)$

- $\left(\int f(x)dx \right)' = f(x)$

- $\int kf(x)dx = k \int f(x)dx$

- $\int [f(x) + g(x)]dx = \int f(x)dx + \int g(x)dx$

- $\int [f(x) + g(x)]dx = \int f(x)dx - \int g(x)dx$

- $\int f(ax)dx = \frac{1}{a}F(ax) + C$

- $\int f(ax + b)dx = \frac{1}{a}F(ax + b) + C$

- $\int f(x)f'(x)dx = \frac{1}{a}f^2(x) + C$

- $\int \frac{f'(x)}{f(x)}dx = \ln|f(x)| + C$

- Method of substitution:

$$\int f(x)dx = f(u(t))u'(t)dt, \quad \text{if } x = u(t)$$

- Integration by parts: $\int udv = uv - \int vdu$, where $u(x)$, $v(x)$ are differentiable functions.

5.8   Integrals of Rational Function

- $\int adx = ax + C$

- $\int xdx = \dfrac{x^2}{2} + C$

- $\int x^2dx = \dfrac{x^3}{3} + C$

- $\int x^pdx = \dfrac{x^{p+1}}{p+1} + C, \quad p \neq -1$

- $\int (ax+b)^ndx = \dfrac{(ax+b)^{n+1}}{a(n+1)} + C, \quad n \neq -1$

- $\int \dfrac{dx}{x} = \ln|x| + C$

- $\int \dfrac{dx}{ax+b} = \dfrac{1}{a}\ln|ax+b| + C$

- $\int \dfrac{ax+b}{cx+d}dx = \dfrac{a}{c}x + \dfrac{bc-ad}{c^2}\ln|cx+d| + C$

- $\int \dfrac{dx}{(x+a)(x+b)} = \dfrac{1}{a-b}\ln\left|\dfrac{x+b}{x+a}\right| + C, \quad a \neq b$

- $\int \dfrac{xdx}{a+bx} = \dfrac{1}{b^2}(a+bx-a\ln|a+bx|) + C$

- $\int \dfrac{x^2dx}{a+bx} = \dfrac{1}{b^3}\left[\dfrac{1}{2}(a+bx)^2 - 2a(a+bx) + a^2\ln|a+bx|\right] + C$

- $\int \dfrac{dx}{x(a+bx)} = \dfrac{1}{a}\ln\left|\dfrac{a+bx}{x}\right| + C$

- $\int \dfrac{dx}{x^2(a+bx)} = -\dfrac{1}{ax} + \dfrac{b}{a^2}\ln\left|\dfrac{a+bx}{x}\right| + C$

- $\int \dfrac{x\,dx}{(a+bx)^2} = \dfrac{1}{b^2}\left(\ln|a+bx| + \dfrac{a}{a+bx}\right) + C$

- $\int \dfrac{x\,dx}{(a+bx)^2} = \dfrac{1}{b^2}\left(\ln|a+bx| + \dfrac{a}{a+bx}\right) + C$

- $\int \dfrac{dx}{x(a+bx)^2} = \dfrac{1}{a(a+bx)} + \dfrac{1}{a^2}\ln\left|\dfrac{a+bx}{x}\right| + C$

- $\int \dfrac{dx}{x^2-1} = \dfrac{1}{2}\ln\left|\dfrac{x-1}{x+1}\right| + C$

- $\int \dfrac{dx}{1-x^2} = \dfrac{1}{2}\ln\left|\dfrac{1-x}{1-x}\right| + C$

- $\int \dfrac{dx}{a^2-x^2} = \dfrac{1}{2a}\ln\left|\dfrac{a+x}{a-x}\right| + C$

- $\int \dfrac{dx}{x^2-a^2} = \dfrac{1}{2a}\ln\left|\dfrac{1-a}{x-a}\right| + C$

- $\int \dfrac{dx}{1+x^2} = \tan^{-1}x + C$

- $\int \dfrac{dx}{a^2+x^2} = \dfrac{1}{a}\tan^{-1}\dfrac{x}{a} + C$

- $\int \dfrac{x\,dx}{x^2+a^2} = \dfrac{1}{2}\ln(x^2+a^2) + C$

- $\int \dfrac{dx}{a+bx^2} = \dfrac{1}{\sqrt{ab}}\arctan\left(x\sqrt{\dfrac{b}{a}}\right) + C, \quad ab>0$

- $\int \dfrac{x\,dx}{a+bx^2} = \dfrac{1}{2b}\ln\left|x^2+\dfrac{a}{b}\right| + C$

- $\int \dfrac{dx}{x(a+bx^2)} = \dfrac{1}{2a}\ln\left|\dfrac{x^2}{a+bx^2}\right| + C$

- $\int \dfrac{dx}{a^2-b^2x^2} = \dfrac{1}{2ab}\ln\left|\dfrac{a+bx}{a-bx}\right| + C$

- $\int \dfrac{dx}{ax^2+bx+c} = \dfrac{1}{\sqrt{b^2-4ac}}\ln\left|\dfrac{2ax+b-\sqrt{b^2-4ac}}{2ax+b+\sqrt{b^2-4ac}}\right| + C$

- $\int \dfrac{dx}{ax^2+bx+c} = \dfrac{2}{\sqrt{4ac-b^2}}\arctan\dfrac{2ax+b}{\sqrt{4ac-b^2}} + C$

5.9   Integrals of Irrational Function

- $$\int \frac{dx}{\sqrt{ax+b}} = \frac{2}{a}\sqrt{ax+b} + C$$

- $$\int \sqrt{ax+b}\, dx = \frac{2}{3a}(ax+b)^{3/2} + C$$

- $$\int \frac{xdx}{\sqrt{ax+b}} = \frac{2(ax-2b)}{3a^2}\sqrt{ax+b} + C$$

- $$\int x\sqrt{ax+b}\, dx = \frac{2(3ax-2b)}{15a^2}(ax+b)^{3/2} + C$$

- $$\int \frac{dx}{(x+c)\sqrt{ax+b}} = \frac{1}{\sqrt{b+ac}}\ln\left|\frac{\sqrt{ax+b}-\sqrt{b-ac}}{\sqrt{ax+b}+\sqrt{b-ac}}\right| + C$$

- $$\int \frac{dx}{(x+c)\sqrt{ax+b}} = \frac{1}{\sqrt{ac+b}}\arctan\left|\frac{\sqrt{ax+b}}{\sqrt{ax-b}}\right| + C$$

- $$\int \sqrt{\frac{ax+b}{cx+d}}\,dx = \frac{1}{c}\sqrt{(ax+b)(cx+d)} - \frac{ad-bc}{c\sqrt{ac}}\ln\left|\sqrt{a(cx+d)}+\sqrt{c(ax+b)}\right| + C, \quad a>0$$

- $$\int \sqrt{\frac{ax+b}{cx+d}}\,dx = \frac{1}{c}\sqrt{(ax+b)(cx+d)} - \frac{ad-bc}{c\sqrt{ac}}\arctan\sqrt{\frac{a(cx+d)}{c(ax+b)}} + C, \quad a<0,\ c>0$$

- $$\int x^2\sqrt{a+bx}\, dx = \frac{2(8a^2-12abx+15b^2x^2)}{105b^3}\sqrt{(a+bx)^3} + C$$

- $$\int \frac{x^2dx}{\sqrt{a+bx}} = \frac{2(8a^2-4abx+3b^2x^2)}{15b^3}\sqrt{a+bx} + C$$

- $$\int \frac{dx}{x\sqrt{a+bx}} = \frac{1}{\sqrt{a}}\ln\left|\frac{\sqrt{a+bx}-\sqrt{a}}{\sqrt{a+bx}+\sqrt{a}}\right| + C, \quad a>0$$

- $$\int \frac{dx}{x\sqrt{a+bx}} = \frac{2}{\sqrt{-a}}\arctan\left|\frac{a+bx}{-a}\right| + C, \quad a<0$$

- $$\int \sqrt{\frac{a-x}{b+x}}\,dx = \sqrt{(a-x)(b+x)} + (a+b)\arcsin\sqrt{\frac{x+b}{a+b}} + C$$

- $$\int \sqrt{\frac{a+x}{b-x}}\,dx = -\sqrt{(a+x)(b-x)} - (a+b)\arcsin\sqrt{\frac{b-x}{a+b}} + C$$

- $$\int \sqrt{\frac{1+x}{1-x}}\,dx = -\sqrt{1-x^2} + \arcsin x + C$$

- $\displaystyle \int \frac{dx}{\sqrt{(x-a)(b-a)}} = 2\arcsin\sqrt{\frac{x-a}{b-a}} + C$

- $\displaystyle \int \sqrt{a + dx - cx^2}\, dx = \frac{2cx - b}{4c}\sqrt{a + bx - cx^2} + \frac{b^2 - 4ac}{8\sqrt{c^3}}\arcsin\frac{2cx - b}{\sqrt{b^2 + 4ac}} + C$

- $\displaystyle \int \frac{dx}{\sqrt{ax^2 + bx + c}} = \frac{1}{\sqrt{a}}\ln\left|2ax + b + 2\sqrt{a(ax^2 + bx + c)}\right| + C, \quad a > 0$

- $\displaystyle \int \frac{dx}{\sqrt{ax^2 + bx + c}} = -\frac{1}{\sqrt{a}}\arcsin\frac{2ax + b}{4a}\sqrt{b^2 - 4ac} + C, \quad a < 0$

- $\displaystyle \int \sqrt{x^2 + a^2}\, dx = \frac{x}{2}\sqrt{x^2 + a^2} + \frac{a^2}{2}\ln\left|x + \sqrt{x^2 + a^2}\right| + C$

- $\displaystyle \int x\sqrt{x^2 + a^2}\, dx = \frac{1}{3}(x^2 + a^2)^{3/2} + C$

- $\displaystyle \int x^2\sqrt{x^2 + a^2}\, dx = \frac{x}{8}(2x^2 + a^2)\sqrt{x^2 + a^2} - \frac{a^4}{8}\ln\left|x + \sqrt{x^2 + a^2}\right| + C$

- $\displaystyle \int \sqrt{\frac{x^2 + a^2}{x^2}}\, dx = -\frac{\sqrt{x^2 + a^2}}{x} + \ln\left|x + \sqrt{x^2 + a^2}\right| + C$

- $\displaystyle \int \frac{dx}{\sqrt{x^2 + a^2}} = \ln\left|x + \sqrt{x^2 + a^2}\right| + C$

- $\displaystyle \int \frac{\sqrt{x^2 + a^2}}{x}\, dx = \sqrt{x^2 + a^2} + a\ln\left|\frac{x}{a + \sqrt{x^2 + a^2}}\right| + C$

- $\displaystyle \int \frac{x\,dx}{\sqrt{x^2 + a^2}} = \sqrt{x^2 + a^2} + C$

- $\displaystyle \int \frac{x^2\,dx}{\sqrt{x^2 + a^2}} = \frac{x}{2}\sqrt{x^2 + a^2} - \frac{a^2}{2}\ln\left|x + \sqrt{x^2 + a^2}\right| + C$

- $\displaystyle \int \frac{dx}{x\sqrt{x^2 + a^2}} = \frac{1}{a}\ln\left|\frac{x}{a + \sqrt{x^2 + a^2}}\right| + C$

- $\displaystyle \int \sqrt{x^2 - a^2}\,dx = \frac{x}{2}\sqrt{x^2 - a^2} - \frac{a^2}{2}\ln\left|x + \sqrt{x^2 + a^2}\right| + C$

- $\displaystyle \int x\sqrt{x^2 - a^2}\,dx = \frac{1}{3}(x^2 - a^2)^{3/2} + C$

- $\displaystyle \int \frac{\sqrt{x^2 - a^2}}{x}\, dx = \sqrt{x^2 - a^2} + a\arcsin\frac{a}{x} + C$

- $\displaystyle\int \frac{\sqrt{x^2-a^2}}{x^2}dx = \frac{\sqrt{x^2-a^2}}{x} + \ln\left|x+\sqrt{x^2-a^2}\right| + C$

- $\displaystyle\int \frac{dx}{\sqrt{x^2-a^2}} = \ln\left|x+\sqrt{x^2-a^2}\right| + C$

- $\displaystyle\int \frac{xdx}{\sqrt{x^2-a^2}} = \sqrt{x^2-a^2} + C$

- $\displaystyle\int \frac{x^2dx}{\sqrt{x^2-a^2}} = \frac{x}{2}\sqrt{x^2-a^2} + \frac{a^2}{2}\ln\left|x+\sqrt{x^2-a^2}\right| + C$

- $\displaystyle\int \frac{dx}{x\sqrt{x^2-a^2}} = -\frac{1}{a}\arcsin\frac{a}{x} + C$

- $\displaystyle\int \frac{dx}{(x+a)\sqrt{x^2-a^2}} = \frac{1}{a}\sqrt{\frac{x-a}{x+a}} + C$

- $\displaystyle\int \frac{dx}{(x-a)\sqrt{x^2-a^2}} = \frac{1}{a}\sqrt{\frac{x+a}{x-a}} + C$

- $\displaystyle\int \frac{dx}{x^2\sqrt{x^2-a^2}} = \frac{\sqrt{x^2-a^2}}{a^2x}$

- $\displaystyle\int \frac{dx}{(x^2-a^2)^{3/2}} = -\frac{x}{a^2\sqrt{x^2-a^2}} + C$

- $\displaystyle\int (x^2-a^2)^{3/2}dx = -\frac{x}{8}(2x^2-5a^2)\sqrt{x^2-a^2} + \frac{3a^4}{8}\ln\left|x+\sqrt{x^2-a^2}\right| + C$

- $\displaystyle\int \sqrt{a^2-x^2}dx = \frac{x}{2}\sqrt{a^2-x^2} + \frac{a^2}{2}\arcsin\frac{x}{a} + C$

- $\displaystyle\int x\sqrt{a^2-x^2}dx = -\frac{1}{3}(a^2-x^2)^{3/2} + C$

- $\displaystyle\int x^2\sqrt{a^2-x^2}dx = \frac{x}{8}(2x^2-a^2)\sqrt{a^2-x^2} + \frac{a^4}{8}\arcsin\frac{x}{a} + C$

- $\displaystyle\int \frac{\sqrt{a^2-x^2}}{x}dx = \sqrt{a^2-x^2} + a\ln\left|\frac{x}{a+\sqrt{a^2-x^2}}\right| + C$

- $\displaystyle\int \frac{\sqrt{a^2-x^2}}{x^2}dx = \frac{\sqrt{a^2-x^2}}{x} + \arcsin\frac{x}{a} + C$

- $\displaystyle\int \frac{dx}{\sqrt{1-x^2}} = \arcsin x + C$

- $$\int \frac{dx}{\sqrt{a^2 - x^2}} = \sin \frac{x}{a} + C$$

- $$\int \frac{x dx}{\sqrt{a^2 - x^2}} = \sqrt{a^2 - x^2} + C$$

- $$\int \frac{x^2 dx}{\sqrt{a^2 - x^2}} = -\frac{x}{2}\sqrt{a^2 - x^2} + \frac{a^2}{2}\arcsin\frac{x}{a} + C$$

- $$\int \frac{dx}{(x+a)\sqrt{a^2 - x^2}} = -\frac{1}{2}\sqrt{\frac{a-x}{a+x}} + C$$

- $$\int \frac{dx}{(x-a)\sqrt{a^2 - x^2}} = -\frac{1}{2}\sqrt{\frac{a+x}{a-x}} + C$$

- $$\int \frac{dx}{(x+b)\sqrt{a^2 - x^2}} = \frac{1}{\sqrt{b^2 - a^2}}\arcsin\frac{bx + a^2}{a(x+b)} + C, \quad b > a$$

- $$\int \frac{dx}{(x+b)\sqrt{a^2 - x^2}} = \frac{1}{\sqrt{b^2 - a^2}}\ln\left|\frac{x+b}{\sqrt{a^2 - b^2}\sqrt{a^2 - x^2} + a^2 + bx}\right| + C$$

- $$\int \frac{dx}{x^2\sqrt{a^2 - x^2}} = -\frac{\sqrt{a^2 - x^2}}{a^2 x} + C$$

- $$\int (a^2 - x^2)^{3/2} dx = \frac{x}{8}(5a^2 - 2x^2)\sqrt{a^2 - x^2} + \frac{3a^4}{8}\arcsin\frac{x}{a} + C$$

- $$\int \frac{dx}{(a^2 - x^2)^{3/2}} = -\frac{x}{a^2\sqrt{a^2 - x^2}} + C$$

5.10   Integrals of Trigonometric Functions

- $$\int \sin x \, dx = -\cos x + C$$

- $$\int \cos x \, dx = \sin x + C$$

- $$\int \sin^2 x \, dx = \frac{x}{2} - \frac{1}{4}\sin 2x + C$$

- $$\int \cos^2 x \, dx = \frac{x}{2} + \frac{1}{4}\sin 2x + C$$

- $\int \sin^3 x \, dx = \frac{1}{3}\cos^3 x - \cos x + C = \frac{1}{12}\cos 3x - \frac{3}{4}\cos x + C$

- $\int \cos^3 x \, dx = \sin x - \frac{1}{3}\sin^3 x + C = \frac{1}{12}\sin 3x + \frac{3}{4}\sin x + C$

- $\int \frac{dx}{\sin x} = \int \csc x \, dx = \ln\left|\tan\frac{x}{2}\right| + C$

- $\int \frac{dx}{\cos x} = \int \sec x \, dx = \ln\left|\tan\left(\frac{x}{2} + \frac{\pi}{4}\right)\right| + C$

- $\int \frac{dx}{\sin^2 x} = \int \csc^2 x \, dx = -\cot x + C$

- $\int \frac{dx}{\cos^2 x} = \int \sec^2 x \, dx = \tan x + C$

- $\int \frac{dx}{\sin^3 x} = \int \csc^3 x \, dx = -\frac{\cos x}{2\sin^2 x} + \frac{1}{2}\ln\left|\tan\frac{x}{2}\right| + C$

- $\int \frac{dx}{\cos^3 x} = \int \sec^3 x \, dx = \frac{\sin x}{2\cos^2 x} + \frac{1}{2}\ln\left|\tan\left(\frac{x}{2} + \frac{\pi}{4}\right)\right| + C$

- $\int \sin x \cos x \, dx = -\frac{1}{4}\cos 2x + C$

- $\int \sin^2 x \cos x \, dx = \frac{1}{3}\sin^3 x + C$

- $\int \sin x \cos^2 x \, dx = -\frac{1}{3}\cos^3 x + C$

- $\int \sin^2 x \cos^2 x \, dx = \frac{x}{8} - \frac{1}{32}\sin 4x + C$

- $\int \tan x \, dx = -\ln|\cos x| + C$

- $\int \frac{\sin x}{\cos^2 x} \, dx = \frac{1}{\cos x} + C = \sec x + C$

- $\int \frac{\sin^2 x}{\cos x} \, dx = \ln\left|\tan\left(\frac{x}{2} + \frac{\pi}{4}\right)\right| - \sin x + C$

- $\int \tan^2 x \, dx = \tan x - x + C$

- $\displaystyle\int \cot x \, dx = \ln|\sin x| + C$

- $\displaystyle\int \frac{\cos x}{\sin^2 x} \, dx = -\frac{1}{\sin x} + C = -\csc x + C$

- $\displaystyle\int \frac{\cos^2 x}{\sin x} \, dx = \ln\left|\tan\frac{x}{2}\right| + \cos x + C$

- $\displaystyle\int \cot^2 x \, dx = -\cot x - x + C$

- $\displaystyle\int \frac{dx}{\cos x \sin x} = \ln|\tan x| + C$

- $\displaystyle\int \frac{dx}{\sin^2 x \cos x} = -\frac{1}{\sin x} + \ln\left|\tan\left(\frac{x}{2} + \frac{\pi}{4}\right)\right| + C$

- $\displaystyle\int \frac{dx}{\sin x \cos^2 x} = \frac{1}{\cos x} + \ln\left|\tan\frac{x}{2}\right| + C$

- $\displaystyle\int \frac{dx}{\sin^2 x \cos^2 x} = \tan x - \cot x + C$

- $\displaystyle\int \sin mx \sin nx \, dx = -\frac{\sin(m+n)x}{2(m+n)} + \frac{\sin(m-n)x}{2(m-n)} + C, \quad m^2 \neq n^2$

- $\displaystyle\int \sin mx \cos nx \, dx = -\frac{\cos(m+n)x}{2(m+n)} - \frac{\cos(m-n)x}{2(m-n)} + C, \quad m^2 \neq n^2$

- $\displaystyle\int \cos mx \cos nx \, dx = \frac{\sin(m+n)x}{2(m+n)} + \frac{\sin(m-n)x}{2(m-n)} + C, \quad m^2 \neq n^2$

- $\displaystyle\int \sec x \tan x \, dx = \sec x + C$

- $\displaystyle\int \csc x \cot x \, dx = -\csc x + C$

- $\displaystyle\int \sin x \cos^n x \, dx = -\frac{\cos^{n+1} x}{n+1} + C$

- $\displaystyle\int \sin^n x \cos x \, dx = \frac{\sin^{n+1} x}{n+1} + C$

- $\displaystyle\int \arcsin x \, dx = x \arcsin x + \sqrt{1 - x^2} + C$

- $\int \arccos x \, dx = x \arccos x - \sqrt{1 - x^2} + C$

- $\int \arctan x \, dx = x \arctan x - \dfrac{1}{2} \ln(x^2 + 1) + C$

- $\int \operatorname{arccot} x \, dx = x \operatorname{arccot} x + \dfrac{1}{2} \ln(x^2 + 1) + C$

5.11   Integrals of Hyperbolic Functions

- $\int \sinh x \, dx = \cosh x + C$

- $\int \cosh x \, dx = \sinh x + C$

- $\int \tanh x \, dx = \ln \cosh x + C$

- $\int \coth x \, dx = \ln|\sinh x| + C$

- $\int \operatorname{sech}^2 x \, dx = \tanh x + C$

- $\int \operatorname{csch}^2 x \, dx = -\coth x + C$

- $\int \operatorname{sech} x \tanh x \, dx = -\operatorname{sech} x + C$

- $\int \operatorname{csch} x \coth x \, dx = -\operatorname{csh} x + C$

5.12   Integrals of Exponential and Logarithmic Functions

- $\int e^x \, dx = e^x + C$

- $\int a^x \, dx = \dfrac{a^x}{\ln a} + C$

- $\int e^{ax} \, dx = \dfrac{e^{ax}}{a} + C$

- $\int x\, e^{ax}\, dx = \dfrac{e^{ax}}{a^2}(ax - 1) + C$

- $\int \ln x\, dx = x \ln x - x + C$

- $\int \dfrac{dx}{x \ln x} = \ln|\ln x| + C$

- $\int x^n \ln x\, dx = x^{n+1}\left[\dfrac{\ln x}{n+1} - \dfrac{1}{(n+1)^2}\right] + C$

- $\int e^{ax} \sin bx\, dx = \dfrac{a \sin bx - b \cos bx}{a^2 + b^2}\, e^{ax} + C$

- $\int e^{ax} \cos bx\, dx = \dfrac{a \cos bx - b \sin bx}{a^2 + b^2}\, e^{ax} + C$

5.13   Reduction Formulas Using Integration by Part

- $\int x^n\, e^{mx}\, dx = \dfrac{1}{m} x^n\, e^{mx} - \dfrac{n}{m} \int x^{n-1}\, e^{mx}\, dx$

- $\int \dfrac{e^{mx}}{x^n}\, dx = -\dfrac{e^{mx}}{(n-1)x^{n-1}} + \dfrac{m}{n-1} \int \dfrac{e^{mx}}{x^{n-1}}\, dx, \quad n \neq 1$

- $\int \sinh^n x\, dx = \dfrac{1}{n} \sinh^{n-1} x \cosh x - \dfrac{n-1}{n} \int \sinh^{n-2} x\, dx$

- $\int \dfrac{dx}{\sinh^n x} = -\dfrac{\cosh x}{(n-1)\sinh^{n-1} x} - \dfrac{n-2}{n-1} \int \dfrac{dx}{x \sinh^{n-2} x}, \quad n \neq 1$

- $\int \cosh^n x\, dx = \dfrac{1}{n} \sinh x \cosh^{n-1} x + \dfrac{n-1}{n} \int \cosh^{n-2} x\, dx$

- $\int \dfrac{dx}{\cosh^n x} = -\dfrac{\sinh x}{(n-1)\cosh^{n-1} x} - \dfrac{n-2}{n-1} \int \dfrac{dx}{\cosh^{n-2} x}, \quad n \neq 1$

- $\int \sinh^n x \cosh^m x\, dx = \dfrac{\sinh^{n+1} x \cosh^{m-1} x}{n+m} + \dfrac{m-1}{n+m} \int \sinh^n x \cosh^{m-2} x\, dx$

- $\int \sinh^n x \cosh^m x\, dx = \dfrac{\sinh^{n+1} x \cosh^{m+1} x}{n+m} + \dfrac{n-1}{n+m} \int \sinh^{n-2} x \cosh^m x\, dx$

- $\int \tanh^n x\, dx = -\dfrac{1}{n-1} \tanh^{n-1} x + \int \cosh^{n-2} x\, dx, \quad n \neq 1$

- $\int \coth^n x \, dx = -\dfrac{1}{n-1} \coth^{n-1} x + \int \coth^{n-2} x \, dx, \quad n \neq 1$

- $\int \operatorname{sech}^n x \, dx = \dfrac{\operatorname{sech}^{n-2} x \tanh x}{n-1} + \dfrac{n-2}{n-1} \int \operatorname{sech}^{n-2} x \, dx, \quad n \neq 1$

- $\int \sin^n x \, dx = -\dfrac{1}{n} \sin^{n-1} x \cos x + \dfrac{n-1}{n} \int \sin^{n-2} x \, dx$

- $\int \dfrac{dx}{\sin^n x} = -\dfrac{\cos x}{(n-1)\sin^{n-1} x} + \dfrac{n-2}{n-1} \int \dfrac{dx}{\sin^{n-2} x}, \quad n \neq 1$

- $\int \cos^n x \, dx = \dfrac{1}{n} \sin x \cos^{n-1} x + \dfrac{n-1}{n} \int \cos^{n-2} x \, dx$

- $\int \dfrac{dx}{\cos^n x} = -\dfrac{\sin x}{(n-1)\cos^{n-1} x} + \dfrac{n-2}{n-1} \int \dfrac{dx}{\cos^{n-2} x}, \quad n \neq 1$

- $\int \sin^n x \cos^m x \, dx = \dfrac{\sin^{n+1} x \cos^{m+1} x}{n+m} + \dfrac{m-1}{n+m} \int \sin^n x \cos^{m-2} x \, dx$

- $\int \sin^n x \cos^m x \, dx = -\dfrac{\sin^{n-1} x \cos^{m+1} x}{n+m} + \dfrac{n-1}{n+m} \int \sin^{n-2} x \cos^m x \, dx$

- $\int \tan^n x \, dx = \dfrac{1}{n} \tan^{n-1} x - \int \tan^{n-2} x \, dx, \quad n \neq 1$

- $\int \cot^n x \, dx = -\dfrac{1}{n-1} \cot^{n-1} x - \int \cot^{n-2} x \, dx, \quad n \neq 1$

- $\int \sec^n x \, dx = \dfrac{\sec^{n-2} x \tan x}{n-1} + \dfrac{n-2}{n-1} \int \sec^{n-2} x \, dx, \quad n \neq 1$

- $\int \csc^n x \, dx = -\dfrac{\csc^{n-2} x \cot x}{n-1} + \dfrac{n-2}{n-1} \int \csc^{n-2} x \, dx, \quad n \neq 1$

- $\int x^n \ln^m x \, dx = \dfrac{x^{n+1} \ln^m x}{n+1} - \dfrac{m}{n+1} \int x^n \ln^{m-1}, x \, dx$

- $\int \dfrac{\ln^m x}{x^n} dx = -\dfrac{\ln^m x}{(n-1)x^{n-1}} + \dfrac{m}{n-1} \int \dfrac{\ln^{m-1} x}{x^n} dx, \quad n \neq 1$

- $\int \ln^n x \, dx = x \ln^n x - n \int \ln^{n-1} x \, dx$

- $\int x^n \sinh x \, dx = x^n \cosh x - n \int x^{n-1} \cosh x \, dx$

- $\int x^n \cosh x \, dx = x^n \sinh x - n \int x^{n-1} \sinh x \, dx$

- $\int x^n \sin x \, dx = -x^n \cos x + n \int x^{n-1} \cos x \, dx$

- $\int x^n \cos x \, dx = x^n \sin x - n \int x^{n-1} \sin x \, dx$

- $\int x^n \sin^{-1} x \, dx = \dfrac{x^{n+1}}{n+1} \sin^{-1} x - \dfrac{1}{n+1} \int \dfrac{x^{n+1}}{\sqrt{1-x^2}} \, dx$

- $\int x^n \cos^{-1} x \, dx = \dfrac{x^{n+1}}{n+1} \cos^{-1} x + \dfrac{1}{n+1} \int \dfrac{x^{n+1}}{\sqrt{1-x^2}} \, dx$

- $\int x^n \tan^{-1} x \, dx = \dfrac{x^{n+1}}{n+1} \tan^{-1} x - \dfrac{1}{n+1} \int \dfrac{x^{n+1}}{\sqrt{1-x^2}} \, dx$

- $\int \dfrac{x^n dx}{ax^n + b} = \dfrac{x}{a} - \dfrac{b}{a} \int \dfrac{dx}{ax^n + b}$

- $\int \dfrac{dx}{(ax^2+bx+c)^n} = \dfrac{-2ax-b}{(n-1)(b^2-4ac)(ax^2+bx+c)^{n-1}}$

  $\qquad\qquad - \dfrac{2(2n-3)a}{(n-1)(b^2-4ac)} \int \dfrac{dx}{(ax^2+bx+c)^{n-1}}, \quad n \neq 1$

- $\int \dfrac{dx}{(x^2+a^2)^n} = \dfrac{x}{2(n-1)a^2(x^2+a^2)^{n-1}} + \dfrac{2n-3}{2(n-1)^2} \int \dfrac{dx}{(x^2+a^2)^{n-1}}, \quad n \neq 1$

- $\int \dfrac{dx}{(x^2+a^2)^n} = \dfrac{x}{2(n-1)a^2(x^2+a^2)^{n-1}} - \dfrac{2n-3}{2(n-1)a^2} \int \dfrac{dx}{(x^2-a^2)^{n-1}}, \quad n \neq 1$

5.14 Definite Integral

Definite integral of a function:

$$\int_a^b f(x)dx, \quad \int_a^b g(x)dx, \quad \ldots$$

Riemann sum:

$$\sum_{i=1}^n f(\xi_i)\Delta x_i$$

Small changes: Δx_i

Antiderivatives: $F(x)$, $G(x)$

Limits of integrations: a, b, c, d

- $\displaystyle \int_a^b f(x)\mathrm{d}x = \lim_{\substack{n \to \infty \\ \max \Delta x_i \to 0}} \sum_{i=1}^n f(\xi_i)\Delta x_i$

  where $\Delta x_i = x_i - x_{i-1},\ x_{i-1} \le \xi_i \le x_i$

- $\displaystyle \int_a^b 1\,\mathrm{d}x = b - a$

- $\displaystyle \int_a^b kf(x)\mathrm{d}x = k \int_a^b f(x)\mathrm{d}x$

- $\displaystyle \int_a^b [f(x) + g(x)]\mathrm{d}x = \int_a^b f(x)\mathrm{d}x + \int_a^b g(x)\mathrm{d}x$

- $\displaystyle \int_a^b [f(x) - g(x)]\mathrm{d}x = \int_a^b f(x)\mathrm{d}x - \int_a^b g(x)\mathrm{d}x$

- $\displaystyle \int_a^b f(x)\mathrm{d}x = 0$

- $\displaystyle \int_a^b f(x)\mathrm{d}x = - \int_b^a f(x)\mathrm{d}x$

- $\displaystyle \int_a^b f(x)\mathrm{d}x = \int_a^c f(x)\mathrm{d}x + \int_c^b f(x)\mathrm{d}x \quad \text{for } a < c < b$

- $\displaystyle \int_a^b f(x)\mathrm{d}x \ge 0, \quad \text{if } f(x) \ge 0 \text{ on } [a, b]$

- $\displaystyle \int_a^b f(x)\mathrm{d}x \le 0, \quad \text{if } f(x) \le 0 \text{ on } [a, b]$

- Fundamental theorem of calculus:

  $\displaystyle \int_a^b f(x)\mathrm{d}x = F(x)|F(x)|_a^b = F(b) - F(a), \quad \text{if } F'(x) = f(x)$

- Method of substitution:

  If $x = g(t)$, then

  $\displaystyle \int_a^b f(x)\mathrm{d}x = \int_c^d f(g(t))g'(t)\mathrm{d}t$

  where $c = g^{-1}(a), d = g^{-1}(b)$

- Integration by parts:

$$\int_a^b u\,dv = (uv)\big|_a^b - \int_a^b v\,du$$

- Trapezoidal rule:

$$\int_a^b f(x)dx = \frac{b-a}{2n}\left[f(x_0) + f(x_n) + 2\sum_{i=1}^{n-1} f(x_i)\right]$$

- Simpson's rule:

$$\int_a^b f(x)dx = \frac{b-a}{3n}[f(x_0) + 4f(x_1) + 2f(x_2) + 4f(x_3) + + 2f(x_4) + \cdots + 4f(x_{n-1}) + f(x_n)]$$

where

$$x_i = a + \frac{b-a}{n}I, \quad I = 0, 1, 2, \ldots, n$$

- Area under a curve:

$$S = \int_a^b f(x)dx = F(b) - F(a)$$

where $F'(x) = f(x)$.
- Area between two curves:

$$S = \int_a^b [f(x) - g(x)]dx = F(b) - G(b) - F(a) + G(a)$$

where $F'(x) = f(x)$, $G'(x) = g(x)$.

5.15   Improper Integral

- The definite integral $\int_a^b f(x)dx$ is called an improper integral.

  If a or b is infinite, $f(x)$ has one or more points of discontinuity in the interval $[a, b]$.
- If $f(x)$ is a continuous function on $[a, \infty)$, then

$$\int_a^\infty f(x)dx = \lim_{n \to \infty} \int_a^n f(x)dx$$

- If $f(x)$ is a continuous function on $(-\infty, b]$, then

$$\int_{-\infty}^{b} f(x)dx = \lim_{n \to -\infty} \int_{n}^{b} f(x)dx$$

Note: The previous improper integrals are convergent if the limits exist and are finite; otherwise the integrals are divergent.

- $$\int_{-\infty}^{\infty} f(x)dx = \int_{-\infty}^{c} f(x)dx + \int_{c}^{\infty} f(x)dx$$

If for some real number c, both of the integrals in the right side are convergent, then the integral $\int_{-\infty}^{\infty} f(x)dx$ is also convergent; otherwise it is divergent.

- Comparison theorems: Let $f(x)$ and $g(x)$ be continuous functions on the closed interval $[a, \infty)$. Suppose that $0 \le g(x) \le f(x)$ for all x in $[a, \infty)$.

5.16   Continuity of a Function

The concept of a continuous function is that it is a function, whose graph has no break. For this reason, continuous functions are chosen, as far as possible, to model the real world problems. If a function is such that its limiting value at a point equals the functional value at that point, then we say that the function is continuous there.

Definition A function $f(x)$ is said to be *continuous* at a point $x = c$, if the following conditions hold true:

1. $f(x)$ is defined at $x = c$
2. $\lim_{x \to c} f(x)$ exists
3. $\lim_{x \to c} f(x) = f(c)$.

If at least one of these conditions is not satisfied, then the function will be *discontinuous* at $x = c$. We say that a function is continuous on an interval, if it is continuous at each point of that interval.

Examples

1. Let a function be such that $f(x) = x^2 + 1$ for $x < 1$ and $f(x) = x$ for $x \ge 1$. Draw the graph of this function and discuss its continuity at the point $x = 1$.
2. Given the function $f(x) = (x^2 - 4)/(x - 2)$ for $x \ne 2$ and $f(2) = 0$. Decide whether this function is continuous on the interval $[0,4]$. Justify your answer.

5.17   Functions and Graphs

Relation R(x): A *relation* is a correspondence between two sets A and B such that each element of set A corresponds to one or more elements of set B. Set A is called the *domain* of the relation and set B is called the *range* of the relation.

Function f(x): A *function* is a relation such that for each element in the domain, there corresponds *exactly one and only one* element in the range. In other words, a function is a well-defined relation.

The function $f(x) = a_n x^n + a_{n-1} x^{n-1} + a_{n-2} x^{n-2} + \cdots + a_1 x + a_0$ is a *polynomial function* of degree n, where n is a nonnegative integer and $a_0, a_1, a_2, \ldots, a_n$ are real numbers. The domain of every polynomial function is $(-\infty, \infty)$.

Rational function: A *rational function* is a function of the form $f(x) = g(x)/h(x)$, where g and h are polynomial functions such that $h(x) \neq 0$. The domain of a rational function is the set of all real numbers such that $h(x) \neq 0$.

Root function: The function $f(x) = \sqrt[n]{g(x)}$ is a *root function*, where n is a positive integer.

1. If n is *even*, the domain is the solution to the inequality $g(x) \geq 0$.
2. If n is *odd*, the domain is the set of all real numbers for which $g(x)$ is defined.

5.18   Partial Fractions

Partial fraction decomposition of $f(x)/g(x)$

1. If the degree of $f(x)$ is not lower than the degree of $g(x)$, use long division to obtain the proper form.
2. Express $g(x)$ as a product of linear factors $ax + b$ or irreducible quadratic $ax^2 + bx + c$, and collect repeated factors so that $g(x)$ is a product of different factors of the form $(ax+b)^n$ or $(ax^2+bx+c)^n$ for a nonnegative integer n.
3. Apply the following rules.

Case I. Distinct Linear Factors

To each linear factor $ax + b$ occurring once in the denominator of a proper rational fraction, there corresponds a single partial fraction of the form $A/(ax + b)$, where A is a constant to be determined.

Case II. Repeated Linear Factors

To each linear factor $ax + b$ occurring n times in the denominator of a proper rational fraction, there corresponds a sum of n partial fractions of the form

$$\frac{A_1}{ax + b} + \frac{A_2}{(ax+b)^2} + \cdots + \frac{A_n}{(ax+b)^n}$$

where the As are constants to be determined.

Case III. Distinct Quadratic Factors

To each irreducible quadratic factor $ax^2 + bx + c$ occurring once in the denominator of a proper rational fraction, there corresponds a single partial fraction of the form $(Ax + B)/(ax^2 + bx + c)$, where A and B are constants to be determined.

Case IV. Repeated Quadratic Factors

To each irreducible quadratic factor $ax^2 + bx + c$ occurring n times in the denominator of a proper rational fraction, there corresponds a sum of n partial fractions of the form

$$\frac{A_1 x + B_1}{ax^2 + bx + c} + \frac{A_2 x + B_2}{(ax^2 + bx + c)^2} + \cdots + \frac{A_n x + B_n}{(ax^2 + bx + c)^n}$$

where the As and Bs are constants to be determined.

5.19 Properties of Trigonometric Functions

Properties of the Sine Function

The sine graph has the following characteristics:

1. the sine function is *odd* since $\sin(-x) = -\sin x$, that is, symmetrical about the origin
2. the sine function is *continuous*
3. the sine function is *periodic with period 2π* since $\sin x = \sin(x + \pi)$
4. $-1 \leq \sin x \leq 1$, that is, the range is $[-1, 1]$
5. the curve cuts the x-axis at 0, $\pm\pi$, $\pm 2\pi$, $\pm 3\pi$, $\pm 4\pi, \ldots$

Graph of $\sin x$

θ	0	$\dfrac{\pi}{4}$	$\dfrac{\pi}{2}$	$\dfrac{3\pi}{4}$	π	$\dfrac{5\pi}{4}$	$\dfrac{3\pi}{2}$	$\dfrac{7\pi}{4}$	2π
$\sin\theta$	0.0	0.7	1.0	0.7	0.0	-0.7	-1.0	-0.7	0.0

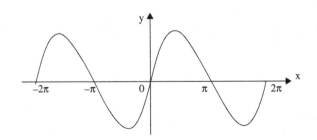

The graph of $y = \dfrac{1}{\sin x} = \text{cosec } x$:

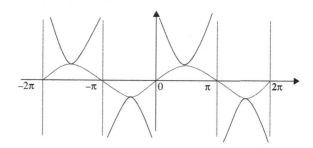

Properties of the Cosine Function

1. The cosine function is *even* since $\cos(-x) = \cos x$, that is, symmetrical about the y-axis
2. The cosine function is *continuous*
3. The cosine function is *periodic with period* 2π since $\cos x = \cos(x + 2\pi)$
4. $-1 \leq \cos x \leq 1$, that is, the range is $[-1, 1]$
5. The curve cuts the x-axis at $0, \pm\dfrac{\pi}{2}, \pm\dfrac{3\pi}{2}, \pm\dfrac{5\pi}{2}, \ldots$

Graph of $\cos x$:

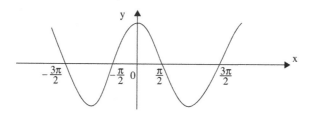

Note: The cosine graph has the same shape as the sine graph but the former is shifted by a distance of $\pi/2$ to the left on the x-axis.

Graph of $y = 1/\cos x = \sec x$:

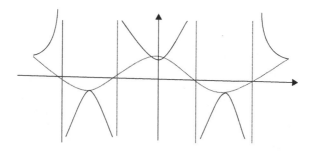

Properties of the Tangent Function

1. The tan function is an *odd* function since $\tan(-x) = -\tan x$

2. The tan function is *not continuous* since $\tan x = \sin x / \cos x$, so $y = \tan x$ is undefined when $\cos x = 0$, that is, when $x = \pm \dfrac{\pi}{2}, \pm \dfrac{3\pi}{2}, \pm \dfrac{5\pi}{2}, \ldots$

These lines $x = \pm \dfrac{\pi}{2}, x = \pm \dfrac{3\pi}{2}, \ldots$ are called *vertical asymptotes*.

The curve approaches these lines but does not touch them.

3. The tangent function is *periodic with period* π since $\tan x = \tan (x + \pi)$

4. $\tan x \in \Re$, that is, range $= \Re$

5. The curve cuts the x-axis at $0, \pm \pi, \pm 2\pi, \ldots$

Graph of tan x:

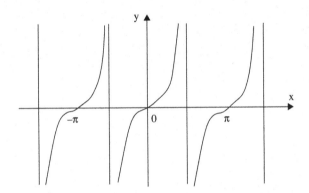

Graph of $y = \dfrac{1}{\tan x} = \cot x$:

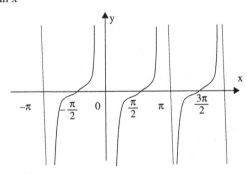

5.20 Sequences and Series

A sequence of numbers is simply a list of numbers, generated by some rule. Examples: 3, 6, 9, 12, . . .

Each term in a sequence can be referred to by its place in the sequence, i.e., first term, third term, nth term. In the examples above: the third term is 9.

A sequence can develop in four ways

Divergent	1, 4, 9, 16, ...	The terms keep growing
Convergent	$4, 2, 1, \frac{1}{2}, \frac{1}{4}, \ldots$	The terms converge on a single value, in this case 0.
Periodic	2, 0, −2, 0, 2, 0, −2, 0, ...	The sequence repeats itself after a set number of terms.
Oscillating	2, −2, −2, 2, −2, ...	The sequence oscillates between two values.

If you add the terms of a sequence together, you get a *series*: $3 + 6 + 9 + 12$.

5.21   Arithmetic Sequences and Series

Arithmetic sequence: Sequences of numbers that follow a pattern of adding a fixed number from one term to the next are called arithmetic sequences. A sequence with general term $a_{n+1} = a_n + d$ is called an *arithmetic sequence, $a_n = n$th term and d = common difference.*

Examples The general (nth) term for 2, 6, 10, 14, 18, 22, ... is 4 and the first term is 2. If we let $d = 4$ this becomes $a_n = a_1 + (n − 1)d$.

The nth or general term of an arithmetic sequence is given by $a_n = a_1 + (n − 1)d$. So in our example $a_1 = 2$ and $d = 4$ so $a_n = 2 + (n − 1)4 = 2 + 4n − 4 = 4n − 2$.

The arithmetic series

To calculate the arithmetic series, we can use $a_n = a_1 + (n − 1)d$, and the sum of the sequence is

$$S_n = \frac{n}{2}(a_1 + a_n) = \frac{n}{2}[2a + (n − 1)d]$$

Examples Find the following sum $3 + 7 + 11 + 15 + \cdots + 35$. We have $a_1 = 3$, $a_n = 35$, $d = 4$. To find n, we note that $35 = 3 + (n − 1)4$ so that $32 = (n − 1)4$ and $n = 9$. Now we are ready to use the formula:

$$S_n = \frac{9}{2}(3 + 35) = 171$$

5.22   Geometric Sequences and Series

Geometric Sequence

Sequences of numbers that follow a pattern of multiplying a fixed number from one term to the next are called geometric sequences. A sequence with general term $a_{n+1} = a_n r$ is called an *geometric sequence*, $a_n = nth$ *term and* $r = common$ *ratio.*

The nth or general term of an geometric sequence is given by $a_n = ar^{n-1}$, where a is the first term.

Examples The general (nth) term for 2, 6, 18, 54, ... is $a_n = a_1(3)^{n-1}$ and the first term is 2.

5.23   The Finite Geometric Series

If $a_n = ar^{n-1}$ is a geometric sequence then the sum of the sequence is

$$S_n = \sum_{i=1}^{n} a_n = a\left(\frac{1 - r^n}{1 - r}\right)$$

Example $3 + 6 + 12 = 21$. This is a geometric series with common ratio 2, where $a = 3$, $a_n = 12$, $r = 2$.

$$S_n = 3\left(\frac{1 - 2^3}{1 - 2}\right) = 3\frac{-7}{-1} = 3(7) = 21$$

5.24   The Infinite Geometric Series

If $a_n = ar^{n-1}$ is a geometric sequence and $|r| < 1$ then the sum of the infinite sequence is

$$S_n = \sum_{i=1}^{\infty} a_n = \frac{a}{1 - r}$$

Examples

- $-2 + 4 - 8 + 16 + \cdots \rightarrow$ We have $a = -2$, $r = -2$ this infinite series diverges because $r = -2$ and $|-2|$ is not < 1. There is no sum.
- $24 + 12 + 6 + 3 + 3/2 + \frac{3}{4} + \cdots \rightarrow$ We have $a = 24$, $r = 1/2$ so that

$$S_n = \frac{24}{1 - (1/2)} = 48$$

5.25 Some of Finite and Infinite Series

$$1 + 2 + 3 + \cdots + n = \frac{n(n+1)}{2}$$

$$2 + 4 + 6 + \cdots + 2n = n(n+1)$$

$$1 + 3 + 5 + \cdots + (2n-1) = n^2$$

$$k + (k+1) + (k+2) + \cdots + (k+n-1) = \frac{n(2k+n-1)}{2}$$

$$1^2 + 2^2 + 3^2 + \cdots + n^2 = \frac{n(n+1)(2n+1)}{6}$$

$$1^3 + 2^3 + 3^3 + \cdots + n^3 = \left(\frac{n(n+1)}{2}\right)^2$$

$$1^2 + 3^2 + 5^2 + \cdots + (2n-1)^2 = \frac{n(4n^2-1)}{3}$$

$$1^3 + 3^3 + 5^3 + \cdots + (2n-1)^3 = n^2(2n^2-1)$$

$$1 + \frac{1}{2} + \frac{1}{2^2} + \frac{1}{2^3} + \cdots + \frac{1}{2^n} + \cdots = 2$$

$$1 + \frac{1}{1!} + \frac{1}{2!} + \frac{1}{2!} + \cdots + \frac{1}{(n-1)!} + \cdots = e$$

5.26 Convergence Tests for Series

A series converges iff the associated sequence of partial sums represented by $\{S_k\}$ converges. The element S_k in the sequence above is defined as the sum of the first "k" terms of the series.

5.27 Series Tests

In this section the various tests mentioned in the previous section will be introduced, and a number of examples will be considered in class to illustrate the various tests.

General (nth) term test (also known as the divergence test):

If $\lim\limits_{n \to \infty} a_n \neq 0$, then the series $\sum_{n=1}^{\infty} a_n$ diverges.

Note This test is a test for divergence only, and says nothing about convergence.

Geometric Series Test

A geometric series has the form $\sum_{n=0}^{\infty} ar^n$, where "a" is some fixed scalar (real number). A series of this type will converge provided that $|r| < 1$, and the sum is $a/(1 - r)$. A proof of this result follows.

Consider the kth partial sum, and "r" times the kth partial sum of the series

$$S_k = a + ar^1 + ar^2 + ar^3 + \cdots + ar^k$$

$$rS_k = ar^1 + ar^2 + ar^3 + \cdots + ar^k + ar^{k+1}$$

The difference between rS_k and S_k is $(r - 1)S_k = a(r^{k+1} - 1)$.

Provided that $r \neq 1$, we can divide by $(r - 1)$ to obtain

$$S_k = \frac{a(r^{k+1} - 1)}{(r - 1)}.$$

Since the only place that "k" appears on the right in this last equation is in the numerator, the limit of the sequence of partial sums $\{S_k\}$ will exist iff the limit as $k \to \infty$ exists as a finite number. This is possible iff $|r| < 1$, and if this is true then the limit value of the sequence of partial sums, and hence the sum of the series, is $S = a/(1 - r)$.

5.28   Integral Test

Given a series of the form $\sum_{n=k}^{\infty} a_n$, set $a_n = f(n)$ where $f(x)$ is a continuous function with positive values that are decreasing for $x \geq k$. If the improper integral

$$\lim\limits_{L \to \infty} \int_{x=k}^{L} f(x)\mathrm{d}x$$ exists as a finite real number, then the given series converges. If the improper integral above does not have a finite value, then the series above diverges.

If the improper integral exists, then the following inequality is always true

$$\int_{x=p+1}^{\infty} f(x)\mathrm{d}x \leq \sum_{n=p}^{\infty} a_n \leq a_p + \int_{x=p}^{\infty} f(x)\mathrm{d}x$$

By adding the terms from $n = k$ to $n = p$ to each expression in the inequalities above it is possible to put both upper and lower bounds on the sum of the series. Also it is possible to estimate the error generated in estimating the sum of the series by using only the first "p" terms. If the error is represented by R_p, then it follows that

$$\int_{x=p+1}^{\infty} f(x)dx \leq R_p \leq \int_{x=p}^{\infty} f(x)dx$$

5.29   Comparison Tests

There are four comparison tests that are used to test series. There are two convergence tests and two divergence tests. In order to use these tests it is necessary to know a number of convergent series and a number of divergent series. For the tests that follow we shall assume that $\sum_{n=1}^{\infty} c_n$ is some known convergent series, that $\sum_{n=1}^{\infty} d_n$ is some known divergent series, and that $\sum_{n=1}^{\infty} a_n$ is the series to be tested. Also it is to be assumed that for $n \in \{1, 2, 3, \ldots, k - 1\}$ the values of a_n are finite and that each of the series contains only positive terms.

5.30   Ratio Test

Given a series $\sum_{n=1}^{\infty} a_n$ with no restriction on the values of the a_ns except that they are finite, and that $\lim\limits_{n \to \infty} |\frac{a_{n+1}}{a_n}| = L$, the series converges absolutely whenever $0 \leq L < 1$, diverges whenever $1 < L \leq \infty$, and the test fails if $L = 1$.

5.31   Absolute and Conditional Convergence

A convergent series that contains an infinite number of both negative and positive terms must be tested for absolute convergence.

A series of the form $\sum_{n=1}^{\infty} a_n$ is *absolutely convergent* iff $\sum_{n=1}^{\infty} |a_n|$ the series of absolute values is convergent.

If $\sum_{n=1}^{\infty} a_n$ is convergent, but $\sum_{n=1}^{\infty} |a_n|$ the series of absolute values is divergent, then the series $\sum_{n=1}^{\infty} a_n$ is *conditionally convergent*.

A shortcut:

In some cases it is easier to show that $\sum_{n=1}^{\infty} |a_n|$ is convergent.

It then follows immediately that the original series $\sum_{n=1}^{\infty} a_n$ is absolutely convergent.

5.32   Taylor and Maclaurin Series

If

$$f(x) = \sum_{n=0}^{\infty} c_n (x-a)^n$$

has a power series representation, then

$$c_n = \frac{f^n(x)}{n!}$$

and

$$f(x) = \sum_{n=0}^{\infty} \frac{f^n(x)}{n!} (x-a)^n = f(a) + f'(a)(x-a) + \frac{f''(a)}{2!}(x-a)^2$$
$$+ \frac{f'''(a)}{3!}(x-a)^3 + \cdots + \frac{f^n(a)}{n!}(x-a)^n + \cdots$$

which is called a *Taylor series* at $x = a$.
   If $a = 0$, then

$$f(x) = \sum_{n=0}^{\infty} \frac{f^n(x)}{n!} x^n = f(0) + f'(0)x + \frac{f''(0)}{2!}x^2 + \frac{f'''(0)}{3!}x^3 + \cdots + \frac{f^n(0)}{n!}x^n + \cdots$$

is the Maclaurin series of f centered at $x = 0$.
   Expansions for some function:

$$\frac{1}{x} = 1 - (x-1) + (x-1)^2 - (x-1)^3 + (x-1)^4 - \cdots + (-1)^n(x-1)^n$$

$$\frac{1}{1-x} = \sum_{n=0}^{\infty} x^n = 1 + x + x^2 + x^3 + x^4 + \cdots$$

$$e^x = \sum_{n=0}^{\infty} \frac{x^n}{n!} = 1 + \frac{x}{1!} + \frac{x^2}{2!} + \frac{x^3}{3!} + \frac{x^4}{4!} + \cdots$$

$$\sin x = \sum_{n=0}^{\infty} (-1)^n \frac{x^{2n+1}}{(2n+1)!} = x - \frac{x^3}{3!} + \frac{x^5}{5!} - \frac{x^7}{7!} + \cdots$$

$$\cos x = \sum_{n=0}^{\infty} (-1)^n \frac{x^{2n}}{(2n)!} = 1 - \frac{x^2}{2!} + \frac{x^4}{6!} - \frac{x^8}{8!} + \cdots$$

$$\tan^{-1} x = \sum_{n=0}^{\infty} (-1)^n \frac{x^{2n+1}}{2n+1} = x - \frac{x^3}{3} + \frac{x^5}{5} - \frac{x^7}{7} + \cdots$$

$$\ln(1+x) = \sum_{k=0}^{\infty} (-1)^k \frac{x^{k+1}}{k+1} = x - \frac{x^2}{2} + \frac{x^3}{3} - \frac{x^4}{4} + \cdots$$

$$(1+x)^\alpha = 1 + \sum_{k=1}^{\infty} \frac{\alpha(\alpha-1)\ldots(\alpha-k+1)}{k!} x^k$$

$$= 1 + \alpha x + \frac{\alpha(\alpha-1)}{2!} x^2 + \frac{\alpha(\alpha-1)(\alpha-2)}{3!} x^3 + \cdots$$

5.33   Continuous Fourier Series

For a function with period T, a continuous Fourier series can be expressed as

$$f(t) = a_0 + \sum_{k=1}^{\infty} a_k \cos(kw_0 t) + b_k \sin(kw_0 t)$$

The unknown Fourier coefficients a_0, a_k, and b_k can be computed as

$$a_0 = \left(\frac{1}{T}\right) \int_0^T f(t)\mathrm{d}t$$

Thus, a_0 can be interpreted as the "average" function value between the period interval $[0, T]$.

$$a_k = \left(\frac{2}{T}\right) \int_0^T f(t)\cos(kw_0 t)\mathrm{d}t$$

$$\equiv a_{-k}(\text{hence } a_k \text{ is an "even" function})$$

$$b_k = \left(\frac{2}{T}\right) \int_0^T f(t)\sin(kw_0 t)\mathrm{d}t$$

$$\equiv -b_{-k}(\text{hence } b_k \text{ is an "odd" function})$$

Example 1

Using the continuous Fourier series to approximate the following periodic function ($T = 2\pi$ s) shown in Figure 5.1:

$$f(t) = \begin{cases} t & \text{for } 0 < t \leq \pi \\ \pi & \text{for } \pi \leq t < 2\pi \end{cases}$$

Specifically, find the Fourier coefficients $a_0, a_1, \ldots, a_8$ and $b_1, \ldots, b_8$.

Solution

The unknown Fourier coefficients a_0, a_k, and b_k can be computed based on the following equations:

$$a_0 = \left(\frac{1}{T}\right) \int_0^{2\pi} f(t) dt$$

$$a_0 = \frac{1}{(2\pi)} \times \left\{ \int_0^{\pi} t\, dt + \int_{\pi}^{2\pi} \pi dt \right\}$$

$$a_0 = 2.35619$$

$$a_k = \left(\frac{2}{T}\right) \int_0^{T=2\pi} f(t) \cos(k w_0 t) dt$$

$$a_k = \left(\frac{2}{2\pi}\right) \times \left\{ \int_0^{\pi} t \cos\left(k \times \frac{2\pi}{T} \times t\right) dt + \int_{\pi}^{2\pi} \pi \times \cos\left(k \times \frac{2\pi}{T} \times t\right) dt \right\}$$

$$a_k = \left(\frac{1}{\pi}\right) \times \left\{ \int_0^{\pi} t \cos(kt) dt + \int_{\pi}^{2\pi} \pi \cos(kt) dt \right\}$$

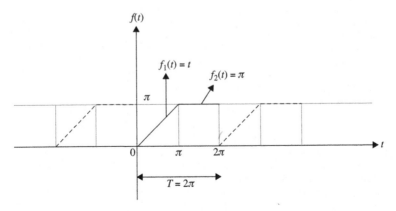

Figure 5.1 A periodic function (between 0 and 2π).

The "integration by part" formula can be utilized to compute the first integral on the right-hand side of the above equation.

For $k = 1, 2, \ldots, 8$, the Fourier coefficients a_k can be computed as

$a_1 = -0.6366257003116296$

$a_2 = -5.070352857678721 \times 10^{-6} \approx 0$

$a_3 = -0.07074100153210318$

$a_4 = -5.070320092569666 \times 10^{-6} \approx 0$

$a_5 = -0.025470225589332522$

$a_6 = -5.070265333302604 \times 10^{-6} \approx 0$

$a_7 = -0.0012997664818977102$

$a_8 = -5.070188612604695 \times 10^{-6} \approx 0$

Similarly

$$b_k = \left(\frac{2}{T}\right) \int_0^{2\pi} f(t) \sin(kw_0 t) dt$$

$$b_k = \left(\frac{1}{\pi}\right) \times \left\{ \int_0^{\pi} t \sin(kt) dt + \int_{\pi}^{2\pi} \pi \sin(kt) dt \right\}$$

For $k = 1, 2, \ldots, 8$, the Fourier coefficients b_k can be computed as

$b_1 = -0.9999986528958207$

$b_2 = -0.4999993232285269$

$b_3 = -0.3333314439509194$

$b_4 = -0.24999804122384547$

$b_5 = -0.19999713794872364$

$b_6 = -0.1666635603759553$

$b_7 = -0.14285324664625462$

$b_8 = -0.12499577981019251$

Any periodic function $f(t)$, such as the one shown in Figure 5.1, can be represented by the Fourier series as

$$f(t) = a_0 + \sum_{k=1}^{\infty} \left\{ a_k \cos(kw_0 t) + b_k \sin(kw_0 t) \right\}$$

where a_0, a_k, and b_k have already been computed (for $k = 1, 2, \ldots, 8$) and

$$w_0 = 2\pi f$$
$$= \frac{2\pi}{T}$$
$$= \frac{2\pi}{2\pi}$$
$$= 1$$

Thus, for $k = 1$, one obtains

$$\bar{f}_1(t) \approx a_0 + a_1 \cos(t) + b_1 \sin(t)$$

For $k = 1 \rightarrow 2$, one obtains

$$\bar{f}_2(t) \approx a_0 + a_1 \cos(t) + b_1 \sin(t) + a_2 \cos(2t) + b_2 \sin(2t)$$

For $k = 1 \rightarrow 4$, one obtains

$$\bar{f}_4(t) \approx a_0 + a_1 \cos(t) + b_1 \sin(t) + a_2 \cos(2t) + b_2 \sin(2t) + a_3 \cos(3t) + b_3 \sin(3t)$$
$$+ a_4 \cos(4t) + b_4 \sin(4t)$$

Plots for $\bar{f}_1(t)$, $\bar{f}_2(t)$, and $\bar{f}_4(t)$ are shown in Figure 5.2.

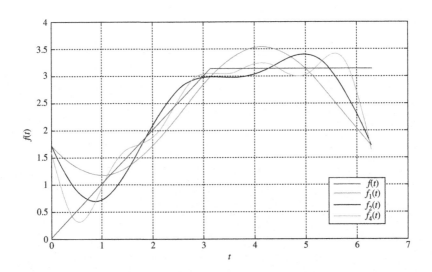

Figure 5.2 Fourier approximated functions.

It can be observed from the figure that as more terms are included in the Fourier series, the approximated Fourier functions more closely resemble the original periodic function as shown in Figure 5.1.

Example 2

The periodic triangular wave function $f(t)$ is defined as

$$f(t) = \begin{cases} \dfrac{-\pi}{2} & \text{for } -\pi < t < \dfrac{-\pi}{2} \\[2mm] -t & \text{for } \dfrac{-\pi}{2} < t < \dfrac{\pi}{2} \\[2mm] \dfrac{-\pi}{2} & \text{for } \dfrac{\pi}{2} < t < \pi \end{cases}$$

Find the Fourier coefficients $a_0, a_1, \ldots, a_8$ and $b_1, \ldots, b_8$ and approximate the periodic triangular wave function by the Fourier series (Figure 5.3).

Solution
The unknown Fourier coefficients a_0, a_k, and b_k can be computed based on the following equations:

$$a_0 = \left(\frac{1}{T}\right)\int_{-\pi}^{\pi} f(t)\,dt$$

$$a_0 = \frac{1}{(2\pi)} \times \left\{ \int_{-\pi}^{-\pi/2}\left(-\frac{\pi}{2}\right)dt + \int_{-\pi/2}^{\pi/2}(-t)\,dt + \int_{\pi/2}^{\pi}\left(-\frac{\pi}{2}\right)dt \right\}$$

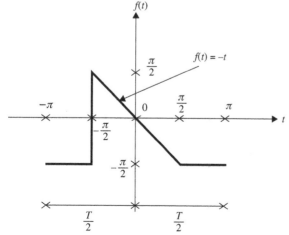

Figure 5.3 Periodic triangular wave function.

$$a_0 = -0.78539753$$

$$a_k = \left(\frac{2}{T}\right)\int_{-\pi}^{\pi} f(t)\cos(kw_0 t)dt$$

where

$$w_0 = \frac{2\pi}{T}$$

$$= \frac{2\pi}{2\pi}$$

$$= 1$$

Hence,

$$a_k = \left(\frac{2}{T}\right)\int_{-\pi}^{\pi} f(t)\cos(kt)dt$$

or

$$a_k = \left(\frac{2}{2\pi}\right)\left\{\int_{-\pi}^{-\pi/2}\left(-\frac{\pi}{2}\right)\cos(kt)dt + \int_{-\pi/2}^{\pi/2}(-t)\cos(kt)dt + \int_{\pi/2}^{\pi}\left(-\frac{\pi}{2}\right)\cos(kt)dt\right\}$$

Similarly

$$b_k = \left(\frac{2}{T}\right)\int_{-\pi}^{\pi} f(t)\sin(kw_0 t)dt = \left(\frac{2}{T}\right)\int_{-\pi}^{\pi} f(t)\sin(kt)dt$$

or

$$b_k = \left(\frac{2}{2\pi}\right)\left\{\int_{-\pi}^{-\pi/2}\left(-\frac{\pi}{2}\right)\sin(kt)dt + \int_{-\pi/2}^{\pi/2}(-t)\sin(kt)dt + \int_{\pi/2}^{\pi}\left(-\frac{\pi}{2}\right)\sin(kt)dt\right\}$$

The "integration by part" formula can be utilized to compute the second integral on the right-hand side of the above equations for a_k and b_k.

For $k = 1, 2, \ldots, 8$, the Fourier coefficients a_k and b_k can be computed and summarized as in Table 5.1.

The periodic function (shown in Example 1) can be approximated by Fourier series as

$$f(t) = a_0 + \sum_{k=1}^{\infty}\left\{a_k \cos(kt) + b_k \sin(kt)\right\}$$

Thus, for $k = 1$, one obtains

$$\bar{f}_1(t) = a_0 + a_1 \cos(t) + b_1 \sin(t)$$

Table 5.1 Fourier Coefficients a_k and b_k for Various k Values

k	a_k	b_k
1	0.999997	−0.63661936
2	0.00	−0.49999932
3	−0.3333355	0.07073466
4	0.00	0.2499980
5	0.1999968	−0.02546389
6	0.00	−0.16666356
7	−0.14285873	0.0126991327
8	0.00	0.12499578

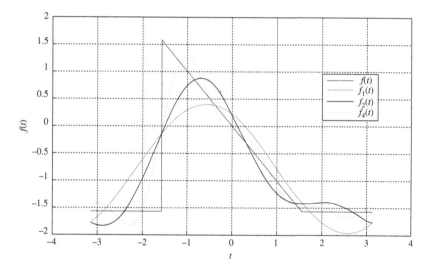

Figure 5.4 Fourier approximated functions.

For $k = 1 \rightarrow 2$, one obtains

$$\bar{f}_2(t) = a_0 + a_1 \cos(t) + b_1 \sin(t) + a_2 \cos(2t) + b_2 \sin(2t)$$

Similarly, for $k = 1 \rightarrow 4$, one has

$$\bar{f}_4(t) = a_0 + a_1 \cos(t) + b_1 \sin(t) + a_2 \cos(2t) + b_2 \sin(2t) + a_3 \cos(3t) + b_3 \sin(3t)$$
$$+ a_4 \cos(4t) + b_4 \sin(4t)$$

Plots for functions $\bar{f}_1(t)$, $\bar{f}_2(t)$, and $\bar{f}_4(t)$ are shown in Figure 5.4.

It can be observed from the figure that as more terms are included in the Fourier series, the approximated Fourier functions closely resemble the original periodic function.

5.34 Double Integrals

A double integral is used to calculate the area under a surface over a bounded region. In order to approximate the volume under a surface over a domain D, the domain can be divided into rectangles. Each of these rectangles has an x and a y dimension denoted as Δx and Δy, respectively. Therefore, the area of each rectangle is defined as $\Delta A = \Delta x \Delta y$.

To obtain the actual volume under a surface, the partitions of the domain must be made infinitely small by finding the infinite limit of the double summations in the volume approximation. As this limit approaches infinity, the error of the approximation approaches 0.

$$
\begin{aligned}
V &= \lim_{m \to \infty} \lim_{n \to \infty} \sum_{i=1}^{m} \sum_{j=1}^{n} f(x_{ij}^*, y_{ij}^*) \Delta A_{ij} \\
&= \lim_{m,n \to \infty} \sum_{i=1}^{m} \sum_{j=1}^{n} f(x_{ij}^*, y_{ij}^*) \Delta A_{ij} \\
&= \iint_{D} f(x, y)\,dA
\end{aligned}
$$

The domain D of the double integral can be broken into the components dx and dy, which produces the notation:

$$
V = \int_{a}^{b} \int_{c}^{d} f(x, y)\,dy\,dx
$$

To evaluate a double integral, the integrand must first be integrated relative to the first differential. All variables other than that of the first differential are treated as constants. The bounds for the inner integral are entered into the antiderivative, which then is integrated relative to the second differential.

For example:

$$
\int_{0}^{2} \int_{1}^{3} (x^2 y)\,dy\,dx
$$

$$
\int_{0}^{2} \left[\frac{1}{2} x^2 y^2 \right]_{1}^{3} dx
$$

$$
\int_{0}^{2} \left(\frac{9}{2} x^2 - \frac{1}{2} x^2 \right) dx
$$

$$
\int_{0}^{2} 4x^2\,dx
$$

$$
\left[\frac{4x^3}{3} \right]_{0}^{2}
$$

$$
\left(\frac{32}{3} - 0 \right)
$$

$$
\frac{32}{3}
$$

5.35 Triple Integrals

A triple integral has a three-dimensional domain. Since the resulting function exists in four dimensions, the function cannot be represented graphically. However, some mathematicians label the results of a triple integral as *hypervolume*. The formula for calculating a triple integral can be determined as

$$\lim_{m \to \infty} \lim_{n \to \infty} \lim_{p \to \infty} \sum_{i=1}^{m} \sum_{j=1}^{n} \sum_{k=1}^{p} f(x_{ijk}^*, y_{ijk}^*, z_{ijk}^*) \Delta z_{jk} \Delta y_{jk} \Delta x_{jk}$$

$$= \iiint_E f(x, y, z) dV$$

$$= \int_a^b \int_c^d \int_e^f f(x, y, z) dz \, dy \, dx$$

Example

Use a triple integral to find the volume of the solid bounded by the graphs of $z = x^2 + y^2$ and the plane $z = 4$.

Solution

The following graph shows a plot of the paraboloid $z = x^2 + y^2$ (in blue), the plane $z = 4$ (in red), and its projection onto the $x - y$ plane (in green).

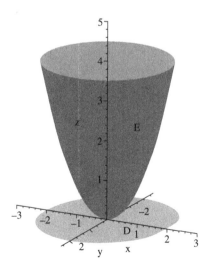

The triple integral $\iiint_E dV$ will evaluate the volume of this surface. In the z direction, the surface E is bounded between the graphs of the paraboloid $z = x^2 + y^2$ and the plane $z = 4$. This will make up the limits of integration in terms of z. The limits for y and x are determined by looking at the projection D given on the $x-y$ plane, which is the graph of the circle $x^2 + y^2 = 4$ given as follows:

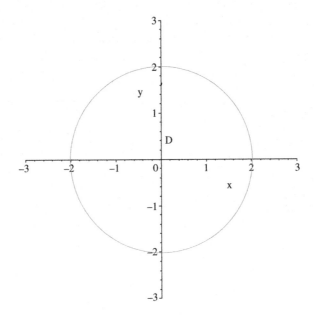

Taking the equation $x^2 + y^2 = 4$ and solving for y gives $y = \pm\sqrt{4-x^2}$. Thus the limits of integration of y will range from $y = -\sqrt{4-x^2}$ to $y = \sqrt{4-x^2}$. The integration limits in terms of x hence range from $x = -2$ to $x = 2$. Thus the volume of the region E can be found by evaluating the following triple integral:

$$\text{Volume of } E = \iiint_E dV = \int_{-2}^{2}\int_{-\sqrt{4-x^2}}^{\sqrt{4-x^2}}\int_{x^2+y^2}^{4} dz\, dy\, dx$$

If we evaluate the innermost integral we get the following:

$$\int_{-2}^{2}\int_{-\sqrt{4-x^2}}^{\sqrt{4-x^2}}\int_{x^2+y^2}^{4} dz\, dy\, dx = \int_{-2}^{2}\int_{-\sqrt{4-x^2}}^{\sqrt{4-x^2}} [z]\Big|_{z=x^2+y^2}^{z=4} dy\, dx$$

$$= \int_{-2}^{2}\int_{-\sqrt{4-x^2}}^{\sqrt{4-x^2}} [4 - (x^2 + y^2)]dy\, dx$$

Since the limits involving y involve two radicals, integrating the rest of this result in rectangular coordinates is a tedious task. However, since the region D on the $x-y$ plane given by $x^2 + y^2 = 4$ is circular, it is natural to represent this region in polar coordinates.

Using the fact that the radius r ranges from $r = 0$ to $r = 2$ and that θ ranges from $\theta = 0$ to $\theta = 2\pi$ and also that in polar coordinates, the conversion equation is $r^2 = x^2 + y^2$, the iterated integral becomes

$$\int_{-2}^{2} \int_{-\sqrt{4-x^2}}^{\sqrt{4-x^2}} [4 - (x^2 + y^2)]dy\, dx = \int_{0}^{2\pi} \int_{0}^{2} (4 - r^2)r\, dr\, d\theta$$

Evaluating this integral in polar coordinates, we obtain

$$\int_{0}^{2\pi} \int_{0}^{2} (4 - r^2)r\, dr\, d\theta = \int_{0}^{2\pi} \int_{0}^{2} (4r - r^3)dr\, d\theta \qquad \text{(distribute } r\text{)}$$

$$= \int_{\theta=0}^{\theta=2\pi} \left(2r^2 - \frac{1}{4}r^4 \Big|_{r=0}^{r=2} \right) d\theta \qquad \text{(integrate)}$$

$$= \int_{\theta=0}^{\theta=2\pi} \left[\left(2(2)^2 - \frac{1}{4}(2)^4 \right) - 0 \right] d\theta \qquad \text{(sub in limits of integration)}$$

$$= \int_{\theta=0}^{\theta=2\pi} 4\, d\theta \qquad \text{(simplify)}$$

$$= 4\theta \Big|_{\theta=0}^{\theta=2\pi} \qquad \text{(integrate)}$$

$$= 4(2\pi) - 4(0) \qquad \text{(sub in limits of integration)}$$

$$= 8\pi$$

Thus, the volume of E is 8π.

5.36   First-Order Differential Equations

Linear equations:

$$\frac{dy}{dx} + g(x)y = f(x)$$

The general solution is

$$y = \frac{\int e^{\int g(x)dx} f(x)dx}{e^{\int g(x)dx}} + C$$

Separable equations:

$$\frac{dy}{dx} = f(x)g(y)$$

The general solution is

$$\int \frac{dy}{g(y)} = \int f(x)dx + C$$

Exact equations: $M(x, y)dx + N(x, y)dy = 0$ is called exact if $\partial M/\partial y = \partial N/\partial x$ and not exact otherwise.
The general solution is

$$\int M(x, y)dx + \int N(x, y)dy = C$$

Homogeneous equations: $dy/dx = f(x, y)$ is homogeneous if the function $f(tx, ty) = f(x, y)$
The substitution $z = y/x$ converts the equation to separable $x(dz/dx) + z = f(1, z)$
Bernoulli equations:

$$\frac{dy}{dx} + g(x)y = f(x)y^n$$

The substitution $z = y^{1-n}$ converts the equation to linear $(dz/dx) + (1 - n)g(x)z = (1 - n)f(x)$

5.37 Second-Order Differential Equations

Homogeneous linear equation with constant coefficients: $y'' + by' + cy = 0$. The characteristic equation is $\lambda^2 + b\lambda + c = 0$.

If $\lambda_1 \neq \lambda_2$ (distinct real roots) then $y = c_1 e^{\lambda_1 x} + c_2 e^{\lambda_2 x}$.
If $\lambda_1 = \lambda_2$ (repeated roots) then $y = c_1 e^{\lambda_1 x} + c_2 x e^{\lambda_1 x}$.
If $\lambda_1 = \alpha + \beta i$ and $\lambda_2 = \alpha - \beta i$ are complex numbers (distinct real roots) then $y = e^{\alpha x}(c_1 \cos \beta x + c_2 \sin \beta x)$.

5.38 Laplace Transform

Why Laplace Transforms?

1. Converts differential equations to algebraic equations—facilitates combination of multiple components in a system to get the total dynamic behavior (through addition and multiplication)
2. Can gain insight from the solution in the transform domain ("s")—inversion of transform not necessarily required

3. Allows development of an analytical model which permits use of a discontinuous (piecewise continuous) forcing function and the use of an integral term in the forcing function (important for control)
4. System analysis using Laplace transform

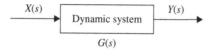

$$Y(s) = G(s) X(s)$$
$$y(t) = L^{-1}(Y(s)) \rightarrow \text{inverse Laplace transform}$$

Definition Laplace transform of $x(t)$

$$L[x(t)] = X(s) = \int_0^\infty x(t)e^{-st}\,dt \quad (s = \sigma + j\omega)$$

5.39 Table of Laplace Transforms

Function, $f(t)$	Laplace Transform, $F(s)$
1	$1/s$
T	$1/s^2$
t^2	$2/s^3$
t^n	$n!/s^{n+1}$
e^{-at}	$1/(s + a)$
$t^n e^{-at}$	$n!/(s + a)^{n+1}$
$\sin(bt)$	$b/(s^2 + b^2)$
$\cos(bt)$	$s/(s^2 + b^2)$
$e^{-at}\sin(bt)$	$b/((s + a)^2 + b^2)$
$e^{-at}\cos(bt)$	$(s + a)/((s + a)^2 + b^2)$
$\sinh(bt)$	$b/(s^2 - b^2)$
$\cosh(bt)$	$s/(s^2 - b^2)$
$t\sin(bt)$	$2bs/(s^2 + b^2)^2$
$t\cos(bt)$	$(s^2 - b^2)/(s^2 + b^2)^2$
$u(t)$ unit step function	$1/s$
$u(t - d)$	e^{-sd}/s
$\delta(t)$	1
$\delta(t - d)$	e^{-sd}

6 Statistics and Probability

Probability and statistics are two related but separate academic disciplines. Statistical analysis often uses probability distributions and the two topics are often studied together. However, probability theory contains much that is of mostly of mathematical interest and not directly relevant to statistics. Moreover, many topics in statistics are independent of probability theory.

Probability (or likelihood) is a measure or estimation of how likely it is that something will happen or that a statement is true. Probabilities are given a value between 0 (0% chance or will not happen) and 1 (100% chance or will happen). The higher the degree of probability, the more likely the event is to happen, or, in a longer series of samples, the greater the number of times such event is expected to happen.

Statistics is the study of the collection, organization, analysis, interpretation, and presentation of data. It deals with all aspects of data, including the planning of data collection in terms of the design of surveys and experiments. Topics discussed in this chapter are as follows:

- Mean
- Median
- Mode
- Standard deviation
- Variance
- Coefficient of variation
- z-Score
- Range
- Central limit theorem
- Counting rule for combinations
- Counting rule for permutations
- Binomial probability
- Poisson probability
- Confidence intervals
- Sample size
- Regression and correlation
- Pearson product−moment correlation coefficient
- Test statistic for hypothesis tests about a population proportion
- Chi-square goodness-of-fit test statistic
- Standard normal distribution table
- Student's t-distribution table
- Chi-square table
- Table of F-statistics, $P = 0.05$

Mathematical Formulas for Industrial and Mechanical Engineering. DOI: http://dx.doi.org/10.1016/B978-0-12-420131-6.00006-3
© 2014 Elsevier Inc. All rights reserved.

6.1 Arithmetic Mean

For sample: $\bar{x} = \frac{\sum x}{n}$. For population: $\mu = \frac{\sum x}{N}$.

6.2 Median

The median is the middle measurement when an odd number (n) of measurement is arranged in order; if n is even, it is the midpoint between the two middle measurements.

6.3 Mode

It is the most frequently occurring measurement in a set.

6.4 Geometric Mean

$$\sqrt[n]{x_1 x_2 \ldots x_n}$$

6.5 Standard Deviation

$$s = \sqrt{\frac{\left(\sum x - \bar{x}\right)^2}{(n-1)}}$$

or

$$s = \sqrt{\frac{n\left(\sum x^2\right) - \left(\sum x\right)^2}{n(n-1)}}$$

6.6 Variance

$$v = s^2$$

6.7 z-Score

$$z = \frac{x - \bar{x}}{s}$$

6.8 Coefficient of Variation

$$CV \ (\%) = \frac{\text{standard deviation}}{\text{mean}} \times 100$$

6.9 Sample Covariance

$$s_{xy} = \frac{\sum (x_i - \bar{x})(y_i - \bar{y})}{n - 1}$$

6.10 Range

Range = largest data value − smallest data value.

6.11 Central Limit Theorem

$$z = \frac{\bar{x} - \mu}{\sigma / \sqrt{n}}, \quad n \geq 30$$

6.12 Counting Rule for Combinations

$$_nC_r = \frac{n!}{r!(n - r)!}$$

6.13 Counting Rule for Permutations

$$_nP_r = \frac{n!}{(n - r)!}$$

6.14 Properties of Probability

Let

P denotes a probability
A, B, C denote specific events

$P(A)$ denotes the probability of event A occurring

$$P(A) = \frac{\text{number of times } A \text{ occurred}}{\text{number of times trial was repeated}}$$

Computing probability using the complement: $P(A) = 1 - P(A^C)$
Addition law: $P(A \cup B) = P(A) + P(B) - P(A \cap B)$
Conditional probability: $P(A|B) = \dfrac{P(A \cap B)}{P(B)}$ or $P(B|A) = \dfrac{P(A \cap B)}{P(A)}$
Multiplication law: $P(A \cap B) = P(B)P(A|B)$ or $P(A \cap B) = P(A)P(B|A)$
Multiplication law for independent events: $P(A \cap B) = P(A)P(B)$
Expected value of a discrete random variable: $E(x) = \mu = \sum xf(x)$

6.15   Binomial Probability Function

$$f(x) = \frac{n!}{x!(n-x)!}p^x(1-p)^{(n-x)}$$

6.16   Expected Value and Variance for the Binomial Distribution

$$E(x) = \mu = np \quad \text{Var}(x) = \sigma^2 = np(1-p)$$

6.17   Poisson Probability Function

$$f(x) = \frac{\mu^x e^{-\mu}}{x!}$$

where $f(x)$ is the probability of x occurrences in an interval, μ is the expected value or mean number of occurrences in an interval, and $e = 2.718$.

6.18   Confidence Intervals

Confidence interval for a mean (large samples): $\bar{x} - z_c\dfrac{\sigma}{\sqrt{n}} < \mu < \bar{x} + z_c\dfrac{\sigma}{\sqrt{n}}$

Confidence interval for a mean (small samples): $\bar{x} - t_c\dfrac{s}{\sqrt{n}} < \mu < \bar{x} + t_c\dfrac{s}{\sqrt{n}}$

$y_p - E < y < y_p + E$, where y_p is the predicted y value for x:

$$E = t_e S_e \sqrt{1 + \dfrac{1}{n} + \dfrac{(x-\bar{x})^2}{SS_x}}$$

Confidence interval for a proportion (where $np > 5$ and $nq > 5$):

$$\dfrac{r}{n} - z_c\sqrt{\dfrac{(r/n)(1-(r/n))}{n}} < p < \dfrac{r}{n} - z_c\sqrt{\dfrac{(r/n)(1-(r/n))}{n}}$$

6.19 Sample Size

Sample size for estimating means $n = \left(\dfrac{z_c \sigma}{E}\right)^2$.

Sample size for estimating proportions $n = p(1-p)\left(\dfrac{z_c}{E}\right)^2$ with preliminary estimate of p.

$n = \dfrac{1}{4}\left(\dfrac{z_c}{E}\right)^2$ with no preliminary estimate of p.

6.20 Regression and Correlation

$$SS_x = \sum x^2 - \dfrac{\left(\sum x\right)^2}{n}$$

$$SS_y = \sum y^2 - \dfrac{\left(\sum y\right)^2}{n}$$

$$SS_{xy} = \sum xy - \dfrac{\left(\sum x\right)\left(\sum y\right)}{n}$$

Least squares line

$y = a + bx$

where $b = SS_{xy}/SS_x$ and $a = \bar{y} - b\bar{x}$.

Standard error of estimate

$$S_e = \sqrt{\dfrac{SS_y - bSS_{xy}}{n-2}}$$

where $b = SS_{xy}/SS_x$.

6.21 Pearson Product–Moment Correlation Coefficient

$$r = \frac{SS_{xy}}{\sqrt{SS_x SS_y}}$$

6.22 Test Statistic for Hypothesis Tests about a Population Proportion

$$Z = \frac{\bar{p} - p_0}{\sqrt{(p_0(1 - p_0))/n}}$$

6.23 Chi-Square Goodness-of-Fit Test Statistic

$$\chi^2 = \sum \frac{(f_o - f_e)^2}{f_e}; \quad df = (c - 1)$$

6.24 Standard Normal Distribution Table

Cumulative probabilities: $P(Z \leq z)$ for $z \leq 0$

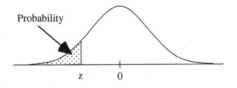

Probability

Z	0.09	0.08	0.07	0.06	0.05	0.04	0.03	0.02	0.01	0.00
−3.5										0.0002
−3.4	0.0002	0.0003	0.0003	0.0003	0.0003	0.0003	0.0003	0.0003	0.0003	0.0003
−3.3	0.0003	0.0004	0.0004	0.0004	0.0004	0.0004	0.0004	0.0005	0.0005	0.0005
−3.2	0.0005	0.0005	0.0005	0.0006	0.0006	0.0006	0.0006	0.0006	0.0007	0.0007
−3.1	0.0007	0.0007	0.0008	0.0008	0.0008	0.0008	0.0009	0.0009	0.0009	0.0010
−3.0	0.0010	0.0010	0.0011	0.0011	0.0011	0.0012	0.0012	0.0013	0.0013	0.0013
−2.9	0.0014	0.0014	0.0015	0.0015	0.0016	0.0016	0.0017	0.0018	0.0018	0.0019
−2.8	0.0019	0.0020	0.0021	0.0021	0.0022	0.0023	0.0023	0.0024	0.0025	0.0026
−2.7	0.0026	0.0027	0.0028	0.0029	0.0030	0.0031	0.0032	0.0033	0.0034	0.0035
−2.6	0.0036	0.0037	0.0038	0.0039	0.0040	0.0041	0.0043	0.0044	0.0045	0.0047
−2.5	0.0048	0.0049	0.0051	0.0052	0.0054	0.0055	0.0057	0.0059	0.0060	0.0062
−2.4	0.0064	0.0066	0.0068	0.0069	0.0071	0.0073	0.0075	0.0078	0.0080	0.0082
−2.3	0.0084	0.0087	0.0089	0.0091	0.0094	0.0096	0.0099	0.0102	0.0104	0.0107
−2.2	0.0110	0.0113	0.0116	0.0119	0.0122	0.0125	0.0129	0.0132	0.0136	0.0139
−2.1	0.0143	0.0146	0.0150	0.0154	0.0158	0.0162	0.0166	0.0170	0.0174	0.0179
−2.0	0.0183	0.0188	0.0192	0.0197	0.0202	0.0207	0.0212	0.0217	0.0222	0.0228
−1.9	0.0233	0.0239	0.0244	0.0250	0.0256	0.0262	0.0268	0.0274	0.0281	0.0287
−1.8	0.0294	0.0301	0.0307	0.0314	0.0322	0.0329	0.0336	0.0344	0.0351	0.0359
−1.7	0.0367	0.0375	0.0384	0.0392	0.0401	0.0409	0.0418	0.0427	0.0436	0.0446
−1.6	0.0455	0.0465	0.0475	0.0485	0.0495	0.0505	0.0516	0.0526	0.0537	0.0548
−1.5	0.0559	0.0571	0.0582	0.0594	0.0606	0.0618	0.0630	0.0643	0.0655	0.0668
−1.4	0.0681	0.0694	0.0708	0.0721	0.0735	0.0749	0.0764	0.0778	0.0793	0.0808
−1.3	0.0823	0.0838	0.0853	0.0869	0.0885	0.0901	0.0918	0.0934	0.0951	0.0968
−1.2	0.0985	0.1003	0.1020	0.1038	0.1056	0.1075	0.1093	0.1112	0.1131	0.1151
−1.1	0.1170	0.1190	0.1210	0.1230	0.1251	0.1271	0.1292	0.1314	0.1335	0.1357
−1.0	0.1379	0.1401	0.1423	0.1446	0.1469	0.1492	0.1515	0.1539	0.1562	0.1587
−0.9	0.1611	0.1635	0.1660	0.1685	0.1711	0.1736	0.1762	0.1788	0.1814	0.1841
−0.8	0.1867	0.1894	0.1922	0.1949	0.1977	0.2005	0.2033	0.2061	0.2090	0.2119
−0.7	0.2148	0.2177	0.2206	0.2236	0.2266	0.2296	0.2327	0.2358	0.2389	0.2420
−0.6	0.2451	0.2483	0.2514	0.2546	0.2578	0.2611	0.2643	0.2676	0.2709	0.2743
−0.5	0.2776	0.2810	0.2843	0.2877	0.2912	0.2946	0.2981	0.3015	0.3050	0.3085
−0.4	0.3121	0.3156	0.3192	0.3228	0.3264	0.3300	0.3336	0.3372	0.3409	0.3446
−0.3	0.3483	0.3520	0.3557	0.3594	0.3632	0.3669	0.3707	0.3745	0.3783	0.3821
−0.2	0.3859	0.3897	0.3936	0.3974	0.4013	0.4052	0.4090	0.4129	0.4168	0.4207
−0.1	0.4247	0.4286	0.4325	0.4364	0.4404	0.4443	0.4483	0.4522	0.4562	0.4602
−0.0	0.4641	0.4681	0.4721	0.4761	0.4801	0.4840	0.4880	0.4920	0.4960	0.5000

Cumulative probabilities: $P(Z \leq z)$ for $z \geq 0$.

z	0.00	0.01	0.02	0.03	0.04	0.05	0.06	0.07	0.08	0.09
0.0	0.5000	0.5040	0.5080	0.5120	0.5160	0.5199	0.5239	0.5279	0.5319	0.5359
0.1	0.5398	0.5438	0.5478	0.5517	0.5557	0.5596	0.5636	0.5675	0.5714	0.5753
0.2	0.5793	0.5832	0.5871	0.5910	0.5948	0.5987	0.6026	0.6064	0.6103	0.6141
0.3	0.6179	0.6217	0.6255	0.6293	0.6331	0.6368	0.6406	0.6443	0.6480	0.6517
0.4	0.6554	0.6591	0.6628	0.6664	0.6700	0.6736	0.6772	0.6808	0.6844	0.6879
0.5	0.6915	0.6950	0.6985	0.7019	0.7054	0.7088	0.7123	0.7157	0.7190	0.7224
0.6	0.7257	0.7291	0.7324	0.7357	0.7389	0.7422	0.7454	0.7486	0.7517	0.7549
0.7	0.7580	0.7611	0.7642	0.7673	0.7704	0.7734	0.7764	0.7794	0.7823	0.7852
0.8	0.7881	0.7910	0.7939	0.7967	0.7995	0.8023	0.8051	0.8078	0.8106	0.8133
0.9	0.8159	0.8186	0.8212	0.8238	0.8264	0.8289	0.8315	0.8340	0.8365	0.8389
1.0	0.8413	0.8438	0.8461	0.8485	0.8508	0.8531	0.8554	0.8577	0.8599	0.8621
1.1	0.8643	0.8665	0.8686	0.8708	0.8729	0.8749	0.8770	0.8790	0.8810	0.8830
1.2	0.8849	0.8869	0.8888	0.8907	0.8925	0.8944	0.8962	0.8980	0.8997	0.9015
1.3	0.9032	0.9049	0.9066	0.9082	0.9099	0.9115	0.9131	0.9147	0.9162	0.9177
1.4	0.9192	0.9207	0.9222	0.9236	0.9251	0.9265	0.9279	0.9292	0.9306	0.9319
1.5	0.9332	0.9345	0.9357	0.9370	0.9382	0.9394	0.9406	0.9418	0.9429	0.9441
1.6	0.9452	0.9463	0.9474	0.9484	0.9495	0.9505	0.9515	0.9525	0.9535	0.9545
1.7	0.9554	0.9564	0.9573	0.9582	0.9591	0.9599	0.9608	0.9616	0.9625	0.9633
1.8	0.9641	0.9649	0.9656	0.9664	0.9671	0.9678	0.9686	0.9693	0.9699	0.9706
1.9	0.9713	0.9719	0.9726	0.9732	0.9738	0.9744	0.9750	0.9756	0.9761	0.9767
2.0	0.9772	0.9778	0.9783	0.9788	0.9793	0.9798	0.9803	0.9808	0.9812	0.9817
2.1	0.9821	0.9826	0.9830	0.9834	0.9838	0.9842	0.9846	0.9850	0.9854	0.9857
2.2	0.9861	0.9864	0.9868	0.9871	0.9875	0.9878	0.9881	0.9884	0.9887	0.9890
2.3	0.9893	0.9896	0.9898	0.9901	0.9904	0.9906	0.9909	0.9911	0.9913	0.9916
2.4	0.9918	0.9920	0.9922	0.9925	0.9927	0.9929	0.9931	0.9932	0.9934	0.9936
2.5	0.9938	0.9940	0.9941	0.9943	0.9945	0.9946	0.9948	0.9949	0.9951	0.9952
2.6	0.9953	0.9955	0.9956	0.9957	0.9959	0.9960	0.9961	0.9962	0.9963	0.9964
2.7	0.9965	0.9966	0.9967	0.9968	0.9969	0.9970	0.9971	0.9972	0.9973	0.9974
2.8	0.9974	0.9975	0.9976	0.9977	0.9977	0.9978	0.9979	0.9979	0.9980	0.9981
2.9	0.9981	0.9982	0.9982	0.9983	0.9984	0.9984	0.9985	0.9985	0.9986	0.9986
3.0	0.9987	0.9987	0.9987	0.9988	0.9988	0.9989	0.9989	0.9989	0.9990	0.9990
3.1	0.9990	0.9991	0.9991	0.9991	0.9992	0.9992	0.9992	0.9992	0.9993	0.9993
3.2	0.9993	0.9993	0.9994	0.9994	0.9994	0.9994	0.9994	0.9995	0.9995	0.9995
3.3	0.9995	0.9995	0.9995	0.9996	0.9996	0.9996	0.9996	0.9996	0.9996	0.9997
3.4	0.9997	0.9997	0.9997	0.9997	0.9997	0.9997	0.9997	0.9997	0.9997	0.9998
3.5	0.9998									

6.25 Table of the Student's *t*-distribution

You must use the *t*-distribution table when working problems when the population standard deviation (σ) is not known and the sample size is small ($n < 30$).

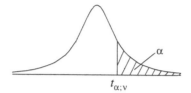

$$t_{\alpha;\nu}$$

The table gives the values of $t_{\alpha;\nu}$, where
$Pr(T_\nu > t_{\alpha;\nu}) = \alpha$ with ν degrees of freedom

ν	α						
	0.1	0.05	0.025	0.01	0.005	0.001	0.0005
1	3.078	6.314	12.076	31.821	63.657	318.310	636.620
2	1.886	2.920	4.303	6.965	9.925	22.326	31.598
3	1.638	2.353	3.182	4.541	5.841	10.213	12.924
4	1.533	2.132	2.776	3.747	4.604	7.173	8.610
5	1.476	2.015	2.571	3.365	4.032	5.893	6.869
6	1.440	1.943	2.447	3.143	3.707	5.208	5.959
7	1.415	1.895	2.365	2.998	3.499	4.785	5.408
8	1.397	1.860	2.306	2.896	3.355	4.501	5.041
9	1.383	1.833	2.262	2.821	3.250	4.297	4.781
10	1.372	1.812	2.228	2.764	3.169	4.144	4.587
11	1.363	1.796	2.201	2.718	3.106	4.025	4.437
12	1.356	1.782	2.179	2.681	3.055	3.930	4.318
13	1.350	1.771	2.160	2.650	3.012	3.852	4.221
14	1.345	1.761	2.145	2.624	2.977	3.787	4.140
15	1.341	1.753	2.131	2.602	2.947	3.733	4.073
16	1.337	1.746	2.120	2.583	2.921	3.686	4.015
17	1.333	1.740	2.110	2.567	2.898	3.646	3.965
18	1.330	1.734	2.101	2.552	2.878	3.610	3.922
19	1.328	1.729	2.093	2.539	2.861	3.579	3.883
20	1.325	1.725	2.086	2.528	2.845	3.552	3.850
21	1.323	1.721	2.080	2.518	2.831	3.527	3.819
22	1.321	1.717	2.074	2.508	2.819	3.505	3.792
23	1.319	1.714	2.069	2.500	2.807	3.485	3.767
24	1.318	1.711	2.064	2.492	2.797	3.467	3.745
25	1.316	1.708	2.060	2.485	2.787	3.450	3.725
26	1.315	1.706	2.056	2.479	2.779	3.435	3.707
27	1.314	1.703	2.052	2.473	2.771	3.421	3.690
28	1.313	1.701	2.048	2.467	2.763	3.408	3.674
29	1.311	1.699	2.045	2.462	2.756	3.396	3.659

v	α						
	0.1	0.05	0.025	0.01	0.005	0.001	0.0005
30	1.310	1.697	2.042	2.457	2.750	3.385	3.646
40	1.303	1.684	2.021	2.423	2.704	3.307	3.551
60	1.296	1.671	2.000	2.390	2.660	3.232	3.460
120	1.289	1.658	1.980	2.358	2.617	3.160	3.373
∞	1.282	1.645	1.960	2.326	2.576	3.090	3.291

6.26   Chi-square Table

df	$P = 0.05$	$P = 0.01$
1	3.84	6.64
2	5.99	9.21
3	7.82	11.35
4	9.49	13.28
5	11.07	15.09
6	12.59	16.81
7	14.07	18.48
8	15.51	20.09
9	16.92	21.67
10	18.31	23.21
11	19.68	24.73
12	21.03	26.22
13	22.36	27.69
14	23.69	29.14
15	25.00	30.58
16	26.30	32.00
17	27.59	33.41
18	28.87	34.81
19	30.14	36.19
20	31.41	37.57
21	32.67	38.93
22	33.92	40.29
23	35.17	41.64
24	36.42	42.98
25	37.65	44.31
26	38.89	45.64
27	40.11	46.96
28	41.34	48.28
29	42.56	49.59
30	43.77	50.89

6.27 Table of F-statistics, P = 0.05

df2	df1																
	1	2	3	4	5	6	7	8	9	10	11	12	13	14	15	20	30
3	10.13	9.55	9.28	9.12	9.01	8.94	8.89	8.85	8.81	8.79	8.76	8.74	8.73	8.71	8.70	8.66	8.62
4	7.71	6.94	6.59	6.39	6.26	6.16	6.09	6.04	6.00	5.96	5.94	5.91	5.89	5.87	5.86	5.80	5.75
5	6.61	5.79	5.41	5.19	5.05	4.95	4.88	4.82	4.77	4.74	4.70	4.68	4.66	4.64	4.62	4.56	4.50
6	5.99	5.14	4.76	4.53	4.39	4.28	4.21	4.15	4.10	4.06	4.03	4.00	3.98	3.96	3.94	3.87	3.81
7	5.59	4.74	4.35	4.12	3.97	3.87	3.79	3.73	3.68	3.64	3.60	3.57	3.55	3.53	3.51	3.44	3.38
8	5.32	4.46	4.07	3.84	3.69	3.58	3.50	3.44	3.39	3.35	3.31	3.28	3.26	3.24	3.22	3.15	3.08
9	5.12	4.26	3.86	3.63	3.48	3.37	3.29	3.23	3.18	3.14	3.10	3.07	3.05	3.03	3.01	2.94	2.85
10	4.96	4.10	3.71	3.48	3.33	3.22	3.14	3.07	3.02	2.98	2.94	2.91	2.89	2.86	2.85	2.77	2.70
11	4.84	3.98	3.59	3.36	3.20	3.09	3.01	2.95	2.90	2.85	2.82	2.79	2.76	2.74	2.72	2.65	2.57
12	4.75	3.89	3.49	3.26	3.11	3.00	2.91	2.85	2.80	2.75	2.72	2.69	2.66	2.64	2.62	2.54	2.47
13	4.67	3.81	3.41	3.18	3.03	2.92	2.83	2.77	2.71	2.67	2.63	2.60	2.58	2.55	2.53	2.46	2.38
14	4.60	3.74	3.34	3.11	2.96	2.85	2.76	2.70	2.65	2.60	2.57	2.53	2.51	2.48	2.46	2.39	2.31
15	4.54	3.68	3.29	3.06	2.90	2.79	2.71	2.64	2.59	2.54	2.51	2.48	2.45	2.42	2.40	2.33	2.25
16	4.49	3.63	3.24	3.01	2.85	2.74	2.66	2.59	2.54	2.49	2.46	2.42	2.40	2.37	2.35	2.28	2.19
17	4.45	3.59	3.20	2.96	2.81	2.70	2.61	2.55	2.49	2.45	2.41	2.38	2.35	2.33	2.31	2.23	2.15
18	4.41	3.55	3.16	2.93	2.77	2.66	2.58	2.51	2.46	2.41	2.37	2.34	2.31	2.29	2.27	2.19	2.11
19	4.38	3.52	3.13	2.90	2.74	2.63	2.54	2.48	2.42	2.38	2.34	2.31	2.28	2.26	2.23	2.16	2.07
20	4.35	3.49	3.10	2.87	2.71	2.60	2.51	2.45	2.39	2.35	2.31	2.28	2.25	2.23	2.20	2.12	2.04
22	4.30	3.44	3.05	2.82	2.66	2.55	2.46	2.40	2.34	2.30	2.26	2.23	2.20	2.17	2.15	2.07	1.98
24	4.26	3.40	3.01	2.78	2.62	2.51	2.42	2.36	2.30	2.25	2.22	2.18	2.15	2.13	2.11	2.03	1.94
26	4.23	3.37	2.98	2.74	2.59	2.47	2.39	2.32	2.27	2.22	2.18	2.15	2.12	2.09	2.07	1.99	1.90
28	4.20	3.34	2.95	2.71	2.56	2.45	2.36	2.29	2.24	2.19	2.15	2.12	2.09	2.06	2.04	1.96	1.87
30	4.17	3.32	2.92	2.69	2.53	2.42	2.33	2.27	2.21	2.16	2.13	2.09	2.06	2.04	2.01	1.93	1.84
35	4.12	3.27	2.87	2.64	2.49	2.37	2.29	2.22	2.16	2.11	2.08	2.04	2.01	1.99	1.96	1.88	1.79

7 Financial Mathematics

The world of finance is literally FULL of mathematical models, formulas, and systems. It is absolutely necessary to understand certain key concepts in order to be successful financially, whether that means saving money for the future or to avoid being a victim of a quick-talking salesman. Financial mathematics is a collection of mathematical techniques that find application in finance, e.g., asset pricing: derivative securities, hedging and risk management, portfolio optimization, structured products. This chapter has links to math lessons about financial topics, such as annuities, savings rates, compound interest, and present value. Topics discussed in this chapter are as follows:

- Percentage
- The number of payments
- Convert interest rate compounding bases
- Effective interest rate
- The future value of a single sum
- The future value with compounding
- The future value of a cash flow series
- The future value of an annuity
- The future value of an annuity due
- The future value of an annuity with compounding
- Monthly payment
- The present value of a single sum
- The present value with compounding
- The present value of a cash flow series
- The present value of an annuity with continuous compounding
- The present value of a growing annuity with continuous compounding
- The net present value of a cash flow series
- Expanded net present value formula
- The present worth cost of a cash flow series
- The present worth revenue of a cash flow series

Symbols used in financial mathematics are as follows:

P: amount borrowed
N: number of periods
B: balance
g: rate of growth
m: compounding frequency

Mathematical Formulas for Industrial and Mechanical Engineering. DOI: http://dx.doi.org/10.1016/B978-0-12-420131-6.00007-5
© 2014 Elsevier Inc. All rights reserved.

r: interest rate
rE: effective interest rate
rN: nominal interest rate
PMT: periodic payment
FV: future value
PV: present value
CF: cash flow
J: the jth period
T: terminal or last period

7.1 Percentage

Percent means "out of one hundred." To change a percent to decimal, drop the % sign, and divide by 100. This is equivalent to moving the decimal point two places to the left. Example: 45%, 76.25%.

7.2 The Number of Payments

$$N = \frac{-\log(1 - rFV/PMT)}{\log(1 + r)}$$

7.3 Convert Interest Rate Compounding Bases

$$r_2 = \left[\left(1 + \frac{r_1}{n_2}\right)^{n_1/n_2} - 1\right]n_2$$

where r_1 is original interest rate with compounding frequency n_1 and r_2 is the stated interest rate with compounding frequency n_2.

7.4 Effective Interest Rate

$$rE = \left\{\left(1 + \frac{rN}{100m}\right)^{[m/(\text{payments/year})]} - 1\right\} \times 100$$

7.5 The Future Value of a Single Sum

$$FV = PV(1 + r)^n$$

7.6   The Future Value with Compounding

$$FV = PV\left(1+\frac{r}{m}\right)^{n-m}$$

7.7   The Future Value of a Cash Flow Series

$$FV = \sum_{j-1}^{n} CF_j(1+r)^i$$

7.8   The Future Value of an Annuity

$$FV_a = PMT\left[\frac{(1+r)^n - 1}{r}\right]$$

7.9   The Future Value of an Annuity Due

$$FV_{ad} = PMT\left[\frac{(1+r)^n - 1}{r}\right](1+r)$$

7.10   The Future Value of an Annuity with Compounding

$$FV_a = PMT\left[\frac{(1+(r/m))^{mn} - 1}{r/m}\right]$$

7.11   Monthly Payment

$$PMT = P\left[\frac{r(1+r)^n}{(1+r)^n - 1}\right]$$

7.12   The Present Value of a Single Sum

$$PV = \frac{FV}{(1+r)^n}$$

7.13   The Present Value with Compounding

$$PV = \frac{FV}{(1+(r/m))^{nm}}$$

7.14   The Present Value of a Cash Flow Series

$$PV = \sum_{j=1}^{n} \frac{FV_j}{(1+(r/m))^{j}}$$

7.15   The Present Value of an Annuity with Continuous Compounding

$$PV_{acp} = \frac{1 - e^{-rt}}{r}$$

7.16   The Present Value of a Growing Annuity with Continuous Compounding

$$PV_{ga} = \frac{PMT(1 - e^{-rt+gt})}{e^{r-g} - 1}$$

7.17   The Net Present Value of a Cash Flow Series

$$NPV = \sum_{j-1}^{n} \frac{CF_j}{(1+r)^{j}}$$

7.18   Expanded Net Present Value Formula

$$NPV = \sum_{T=0}^{T} \frac{CF_T}{(1+r)^{T}}$$

7.19   The Present Worth Cost of a Cash Flow Series

$$\text{PWC} = \sum_{j-1}^{n} \frac{\text{CF}_j}{(1+r)^j}$$

where $\text{CF}_j < 0$.

7.20   The Present Worth Revenue of a Cash Flow Series

$$\text{PWR} = \sum_{j-1}^{n} \frac{\text{CF}_j}{(1+r)^j}$$

where $\text{CF}_j > 0$.

Printed in the United States
By Bookmasters